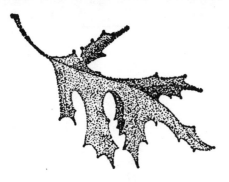

The Wonder of It:

Exploring How the World Works

edited by Bonnie Neugebauer

Published by Child Care Information Exchange/Exchange Press Inc.

Acknowledgments

Production Editor

Sandy Brown

Photographers

Nancy P. Alexander
Toni Liebman
Michael Siluk
Subjects and Predicates
Francis Wardle

Rubber Stamps

Daffodils, snowflake, feather, damsel fish, sea otter, venus comb shell,
dinosaur, zebras, and monarch butterfly copyrighted by and
used with the permission of Susan Manchester
and Imprints Graphic Studio, Inc.—
Graphistamp, Box 3656, Carmel, CA 93921

Raindrop, rabbit, lobster, landing bird, footprint, caterpillar, popcorn,
shuttle, storm cloud, squirrel, ant, leaves, chef hat, goldfish,
spider and web, bubbles, and lightning bolt
copyrighted by and used with the permission of
Rubber Stamps of America, PO Box 567, Saxtons River, VT 05154

Masculine and feminine pronoun references in this book are used randomly for
simplicity and in no way reflect stereotyped concepts of children or adults.

ISBN 0-942702-05-0
Printed in the United States of America

The Wonder of It:
Exploring How the World Works

Table of Contents

A Place to Begin

I thought of you and the science for young children project yesterday when one of my favorite four year olds helped me yank the potato plants out of the window box. We planted them when the potatoes sprouted and now they had reached the ceiling. Josie thought we shouldn't take them out but rather saw a hole in the ceiling for each of the eight plants. But since the storage space over the ceiling has no windows (a fact we ascertained by complicated measuring in and out), he abandoned the idea. We then planted onions and bird seeds (he *hopes* birds will emerge although he knows they won't). Well, all this is to say that science must be active and open to improbable thoughts.

Science is so much a part of everyday life that any alert teacher who believes in **active** learning is surrounded by science opportunities. I've seen people do magic—really involving kids deeply.

Ilse Mattick
Jamaica, Vermont

Thank you, Ilse, for the perfect introduction. Let's talk about science and wonder . . .

Where Has All the Science Gone?

by Jim Greenman

"I do not know what I may appear to the world; but to myself I seem to have been only like a boy playing on the seashore, and diverting myself in now and then finding a smoother pebble or a prettier shell than ordinary, while the great ocean of truth lay all undiscovered before me."

—*Issac Newton in Memoirs of the Life, Writings, and Discoveries of Sir Issac Newton. New York: Johnston Co., 1965.*

Children are natural scientists, impelled to dive into the primordial muck with all their powers and make sense of it all. **Make** sense—taste and twist and rub and bang and hurl the elements against each other and themselves.

The problem is that the natural world of people and things is messy —*mucky* even—and quite dangerous, particularly in the clutches of a curious child. The more enthusiasm, the more inelegant the miniature scientist. And the civilized world, as we like to think of ourselves, is increasingly fastidious and cautious. Indoor malls set the tone: trees flourish without visible water

or dirt and autumn leaves do not fall because there is no autumn in a mall, only Halloween displays and Thanksgiving sales. We are well on our way to the *malling* of childhood. The shadow world is losing ground to fluorescent light. And as children have become increasingly valuable objects of consumption—clothes, classes, coiffures—the tolerance for vigorous mucking about is dropping. An opportunistic insurance industry and a litigious fashion sweeping the culture have banished the notion of accidents from life; there is only negligence and liability.

The child's laboratory is shrinking rapidly in cities and suburbs. Unmanaged experience that children shape and give meaning to through their investigations require some exposure to fields and streams, or vacant lots and rainy streets. The time available for child science and for freedom from the eyes of solicitous and restricting adults shrinks as child care and organized activities fill the child's day.

There are also the laboratories of the kitchen and garage, made obsolete by Betty Crocker and Mr. Goodwrench. Erector and chemistry sets

have been replaced by Nintendo and television. The social science laboratory of block games like "Capture the Flag" has been supplanted by littler and littler Little Leagues.

Let the Sunshine In

If children are to develop a sense of science, if we are to have science in the classrooms, we need to bring life into the classroom and the class out into the world. We must work hard to develop a daily routine that allows children to experience life in gulps and sips as well as measured doses. And life is what? Sunlight and breezes and frost on the win-

Water
by Hilda Conkling

The world turns softly
Not to spill its lakes and rivers.
The water is held in its arms
And the sky is held in the water.
What is water,
That pours silver,
And can hold the sky?

dow pane. Shadows slowly descending on a corner of the room. Water, seizing the light or revealing mysterious caverns in its shadowy reflection. A parakeet chattering away while martins dine at the feeder outside the window. Sifting and mixing and kneading and baking. Pouring and dripping and spilling and measuring. Feeding and nurturing and breeding and burying guinea pigs, mice, rabbits.

Science takes discipline and order—you don't just groove on nature. But it is a discipline to the task, the investigation, not to the 20 minute schedule allotment or to turn-taking. It is an interest-driven discipline most children possess. Witness a baby investigating the archaeological remains of lunch on the floor—systematically in pursuit of the edible. Or a five year old trying to achieve the right trajectory and distance to hit a floating log with a rock. Or an eight year old trying to build a fire.

A child engages in two worlds—the world that she can measure and calibrate and the world that she can feel with her heart and imagination. When children search and question, they are dogged in pursuit of the answer if they have been taken seriously and listened to in the past. But investigations often follow no straight lines. They are made up of bits and pieces, episodic and varied, which call for a different kind of order, following both heart and mind.

Interesting errors and side effects drive both serious science and child explorations toward important learning. The critical issue in many early childhood programs: how to allow decidedly non-linear explorations in decidedly adult oriented linear time and space.

Does Science Require Role Models? Inquiring Minds Want to Know

"The purpose of life, after all, is to live it, to taste experience to the utmost, to reach out eagerly and without fear for newer and richer experience."

—*Eleanor Roosevelt*

In many programs I work with, there are few people on the staff with much interest in the way the world works. In the good schools and programs, there are many people who love being with children and who enjoy the enthusiasm and wonder that children bring to the classroom. They reflect the child's curiosity, but they have little of their own. Why do leaves turn red? Why does popcorn pop? Why is the vacuum not working? Why

Photograph by Francis Wardle

are people in Afghanistan at war? They live in the world pretty much without question. When they ask questions of the children (and they do so) or respond to the child's questions, it is in a gentle teacher-like fashion, as a good early childhood teacher should respond.

Perhaps that is enough for the most part: teachers who care deeply about the children and listen to them. After all, in many child care programs and schools, one is hard pressed to find many teachers having actual conversations with children, particularly conversations driven by the child's interest.

But what a difference it makes in a program to have an enthusiast or two, someone who is driven to ask "why" and "how," just like a child. In those classrooms, science comes alive, much as music pervades the air in the room of a music lover.

The enthusiast may not be the teacher, or even an adult. In one classroom, the adolescent son of the director with an obsession for bug collecting captivated the class and perhaps added to the future supply of entomologists. In another, it was a classroom aide who loved plants. It may be the handyman, a parent, even the principal or director—someone who carries with them a charge of intellectual electricity.

I am concerned for all educational programs, but child care in particular, for it has yet to be solidified into a rigid social form. In a rush to credentialize and professionalize and secure our perimeter, I hope we do not eliminate many of the enthusiasts from the scene. In many programs, a rich source of intellectual life comes from people in their early twenties *passing through* child care—working with children for a while on the way to other lives in the worlds of natural science, music, theater, the environment, academia. Most of the men and non-traditional women who ultimately stay in child

care fit this description. But if the only positions open for these people are as aides working under teachers with two and four year degrees or less, their numbers will continue to plummet.

Come Gather Round Children

There is no doubt a role for science curriculum with group activities and experiments in early childhood. Just as there is an important role for zoos in preserving and developing appreciation for the wildlife of the planet. But there is a sadness about it all when the artificial replaces the natural. Children are born with the secret of the sensuous life in the rich objective world, and childhood science is inescapably loaded with sensuality. Children come to us ready and eager to know and we put them on hold. In **Buckminster Fuller to Children of Earth**, Buckminster Fuller writes, "A child is comprehensive. He wants to understand the whole thing . . . UNIVERSE. Children will draw pictures with everything in them . . . houses and trees and people and animals . . . and the sun AND the moon. Grown-ups say, 'That's a nice picture, Honey, but you put the moon and the sky at the same time and that isn't right.' But the child is right! The sun and the moon ARE in the sky at the same time.

Adults are often busy. They don't answer the child's questions. And the child goes to school and the teacher says, 'First you're going to learn A, B, C. . . .' The child still wants to understand UNIVERSE and has BIG questions, and the teacher says, 'Never mind that . . . you learn the parts first . . . A, B, C. . . .' Then the child goes to college and never does get back to the whole."

First Things First

Why focus on science curriculum when the life is being squeezed out of early childhood? When we have established programs with open windows and teeming life inside, and puddle-rich, flowered playgrounds abuzz with construction outside, when programs have people large and small with an enthusiastic appreciation of nature and the way things work, then let us worry about organized curricular science activities for preschool children.

Reference

Fuller, R. Buckminster. **Buckminster Fuller to Children of Earth.** Garden City, NY: Doubleday, 1972.

Jim Greenman writes about child care and early education from the standpoint of a former teacher, director, college instructor, child care parent, and present day child at heart. Author of **Caring Spaces, Learning Places: Children's Environments That Work** *(Exchange Press) and editor of* **Making Day Care Better: Training, Evaluation, and the Process of Change** *(Teachers College Press), he is currently a child care consultant; director of programs and facilities development for Resources for Child Care Management; and director of* **Project Organizational Quality**—*a three year McKnight Foundation Project to improve centers as businesses operations and to improve the wages, benefits, and working conditions of child care staff.*

Sharing the Wonder—

Teachers and Children Together

Some People
by Rachel Field

Isn't it strange some people make
    You feel so tired inside,
Your thoughts begin to shrivel up
    Like leaves all brown and dried!

But when you're with some other ones,
    It's stranger still to find
Your thoughts as thick as fireflies
    All shiny in your mind!

Kids and Science—
Magic in the Mix

by Sally Cartwright

A valid introduction of science in the early childhood classroom can enliven children's curiosity and their drive to learn from observation, invention, and discovery for the rest of their lives. But this seldom happens, because it requires a process of education shot through with magic. Here's a look at the magic and some ways to achieve it.

One autumn morning in our small, rural preschool here on the Maine coast, Nina brought in a cardboard carton containing a short milkweed stem and leaves.

"Look what I found! A chrysalis! It's very *deckilate*." And so began a science experience, totally un-planned.

Nina, usually awkward and shy, took charge. I was delighted: instead of *hanger-on* she now became our expert. She advised a single layer of cheesecloth be placed across the top of the box. We complied. The gauze deterred probing fingers, allowed sufficient air and light to reach her specimen, and we could see through its airy mesh with ease. The monarch chrysalis, leaf green and flecked with gold, was captivating.

Two children, a parent in tow, went to our village library and found a picture book of the monarch butter-fly's life cycle. I read this aloud to the whole group. Later, in the shelter of the reading corner, the children, quite on their own in twos and threes, continued *reading* the pictures to each other. All avidly discussed various *"callipitters,"* even became them, creeping to *caterpillar music* from my accordion, and curling into pupae at the music's end.

A creative if unscientific rash of eggs, larvae, and pupae appeared spontaneously in clay; a wonderful smudge of colors at the wall easel was called a butterfly; and our tiny green and gold monarch-in-the-making became our friend.

"Can we keep him?" asked Mark.

"No, monarch butterflies go south," Nina answered.

"Like the geese we saw," put in Jan.

"Yup, where it's warm in winter," said Tom.

I wondered where "south" meant, but I didn't elaborate. How impor-tant not to distract these young children with an intrusion of more information than they can use!

Over the next few days, the chil-dren, entirely on their own volition and with mounting interest, watched the chrysalis darken. Distinctive markings developed beneath its pupal shell. One morn-ing they found an orange and black butterfly, huddled and wet, the frail shell split open, still dangling nearby. Without a word from me, the children from all parts of the room put down their paint brushes, left their blocks, their clay, their dramatic play, and wordlessly gathered by the box. This tiny but remarkable metamorphic climax in nature was for us a shared wonder and brought us very close. Mark's hand slipped into Nina's.

Earlier, Lisa had puzzled why the pupa darkened, and I asked her if

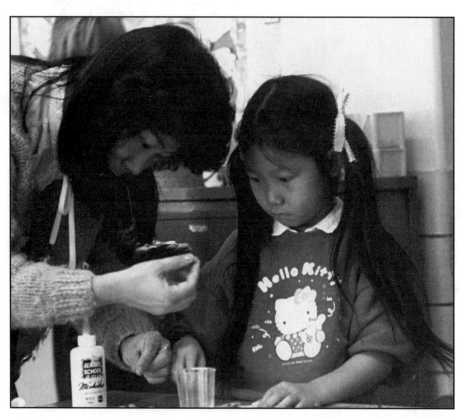

Photograph by Francis Wardle

she might like to check its color with a picture of the adult butterfly. She slipped away with a quiet purpose. Moments later, she startled us, for she sprang up shouting, "Wings! It's wings!" Her sense of discovery was part of our magic, but her shrill exclamation drew four young skeptics to her side. She explained her find to them by again comparing pupa and pictures. (Good science learning means not only understanding but *using* the information to one's own purpose.)

And when the real butterfly emerged, these youngsters knew for sure they'd seen the wings forming in the chrysalis. They laughed and hugged each other. My job as teacher was often to stay clear. Children need to find and feel their competence.

Next morning, when the monarch's wings were dry and strong, we sat around its box on the school porch in hushed excitement. Nina drew back the cheesecloth. Our monarch fluttered up over the playground and zigzagged southward. Almost every child watched until our fragile friend was a tiny speck that vanished in the pale sky. Not a sound came from the kids. Then Mark whispered, "I'm glad he's free."

Part of the magic in this experience was its sense of adventure. I did not want to decide ahead of time what should be taught. I purposely didn't ask the children to learn a thing. Nor did I ask them to verbalize their experience.

My aim was not to give the kids a broadside of information. They each have a lifetime in which to learn facts. I wanted them to *get the feel* of learning through compassionate initiative, participation, observation, and intuitive insight. They had to do this through their own action, not through talking about it.

To be sure, young children are more physical than verbal, but any child (or adult) needs to know this kind of deeply felt, right-brained learning *through direct experience.* In our society, with its left-brained, linear, intellectual rationale dominating our schools, experiential/intuitive learning seems, by comparison, sheer magic. Nevertheless, infants and toddlers depend on it. Both walking and talking are best learned in this way. We teachers need to nurture our children's innate curiosity, their imagination, their creativity, their purposeful, child-initiated work toward discovery.

It's so easy to drown a child's resourceful learning initiative in teacher-directed procedures. Instead, I warmly supported the children with my own genuine interest. But I wouldn't steer them. I trusted these young kids to use their intrinsic curiosity, intuition, and compassion. I trusted them to make connections, to see relationships, to really think, and then to test their findings on their own. This process, using both right and left brain, the yin and yang in each of us, is not only valuable learning—it is the essence of scientific discovery.

A good teacher knows with caring respect not only this wonderful child process of learning, but also that a child develops unevenly, in

spurts, plateaus, and setbacks. Physical, social, and cognitive achievements are interdependent and inseparable from emotional development. For example, as Nina's self-esteem grew, her coordination improved, and sometimes her leadership (social competence) became as deft as her fingers. When the monarch flew off, Nina partially reverted to her somewhat clumsy, uncertain self. It takes a long time to grow. And for teachers it takes our finest knowledge of child development, our patience, our compassion, our detachment, our humor, integrity, and trust.

The nursery child has a wonderful drive and capacity to learn on her own. She finds out what she needs to know through her own effort quite as she learned to walk. There need be no contrived motivation. Learning is hard work. It's often frustrating. At the same time, it can be an exciting challenge, a living, expanding adventure for the children. I've worked hard over the years to quicken this magic in the mix—a child's curiosity, wonder, intuition, purpose, and caring.

Magic for a good science program depends on the setting and spirit of the entire curriculum. A small child needs to feel assured and easy in his surroundings if he is to turn his full energy to building skills and to learning effectively. Thus, along with the steady, warm support of the teachers, the setting and routine

were intentionally simple, consistent, and predictable. We had a large, sunlit classroom overlooking the sea. It was a workshop for the 12 children, and its simple furnishings formed a quiet background for child activity in a child's world. The design and use of space—each piece of equipment, each material, each toy—was carefully chosen to encourage active, resourceful learning, cooperative endeavor, and the child's deep sense of himself as an able learner who was liked and

Photograph by Nancy P. Alexander

valued by his peers. Structured toys were few because they tended to dictate function to the child, whereas the carefully chosen basic materials invited child autonomy, creativity, and—in group work—remarkable cooperation and friendship. It was a joy to be with these children and feel their ingenuity, humor, and compassion for each other as well as for their work of learning.

Science was no exception, and most of it was a natural part of our daily experience. For example, in block building and play, *children learned through experiment* both the perception and use of dimension, weight, balance, levers, and cantilever. At the water table, *they discovered for themselves*, as it served their own purpose, such usually non-verbalized phenomena as floating and sinking, centers of gravity ("what's top heavy tips"); an intuitive sense (with continuing awe and delight) of water qualities—its fluidity, viscosity, transparency, surface tension, and wetness (!); a fantasy world of bubbles, iridescent in sunlight; an incipient feeling for air pressure; and 2 of our 12 youngsters, with remarkable persistence, made and used siphons. Outside, the rock and sand sea edge was an intertidal biology world of endless fascination.

In conclusion, my field notes of a science venture on the edge of the sea (see box) illustrate some of the magic in good child learning.

Sally Cartwright, who began professional work with children over 40 years ago, has taught preschool through junior high children, both as a classroom teacher and as a specialist in elementary science. Her professional writing includes eight science-related books for children. In the fishing village of Tenants Harbor, Maine, where she currently lives, she was founder and teaching director of the Community Nursery School, a seven year experimental unit to develop child initiative for effective learning.

May 8: Drift-in Beach

No one here but us (12 kids, 2 teachers). Sea shining in the sun. Light, warm breeze. Clean, salt smell. Children ran pell-mell over the sand, its level expanse at ebb tide inviting freedom and joyous movement. They explored the rocks, tide pools, boulders. They found tiny animals and shells and seaweeds. They tested their skills—leaping, jumping, climbing, sliding, falling, and dodging pools to stay dry. Ha! Nearly all feet were wet, with Josh in up to his knees and his arms soaking to his elbows, heavy jacket and all. We poured water out of sneakers and boots, but no one was cold.

The children aided each other over the rocks, exploring together, shouting their findings, sharing their joys. A couple of kids were digging in rock-strewn sand when they discovered water rising in their hole.

"Where does this come from?"

"It's impossible. The sea is way down the beach."

"How can water come out of the sand?"

"Come on! Let's find out."

They dug with their hands. They removed heavy rocks, one by one, gradually uncovering a hidden stream. Jason took the lead as other children joined the work. Within 15 minutes, every child was there and working hard. Some rocks took three kids lifting together to budge them.

Jason tasted the water.

"It's fresh!" he shouted. Others checked his assertion. They tasted the spring water and ran down to the sea to taste its salt for comparison. Then they were back at work, intent on the process of discovery. Although they found where the clear, cool spring water welled up through the rocks, that goal seemed unimportant. What caught their hearts (and mine) was their concerted action and spirit. These preschool children excavated—through rock and sand with their small, bare hands—28 feet of hidden, winding stream. They were helping each other, deeply involved, eager, joyous, and determined.

Ardent Curiosity Propels Children Toward Science

by Marguerita Rudolph

Zina and Nina were sitting by a small stream, arguing whether the stream was alive or not.

Zina: "Does it run? It does. And if stopped, does it stand still? It does! Then it's alive."

Nina: "It runs, and it stands—but it can't sit! So it's *not* alive."

Zina: "When I blow on it, it wrinkles up. So it *is* alive."

Nina: "It wrinkles, but doesn't turn away—so it's not *alive*."

Zina: "It's not alive—but seems alive."

The girls agreed on that.

—translated by Marguerita Rudolph from a typed Russian record

It is not uncommon for adults, including teachers, to be skeptical about an appropriate science program in preschool and early school classes. "Real *science* for *young* children?" they ask. "What does it mean?"

Some renowned educators and scientists have considered these questions. The philosopher John Dewey, pondering on the extent and nature of science education at the beginning of the century, stated his "... conviction ... that the motive and unspoiled attitude of childhood, marked by ardent curiosity, fertile imagination and love of experimental inquiry is near, very near to the attitude of the scientific mind" (Dewey, 1919).

Dewey's thinking can be demonstrated by four year old Sheila.

On her own, Sheila notices a raw onion on the nursery school table. After initial silent inspection, she persists playfully in touching, rubbing, peeling with fingernails, blinking, sniffing, grimacing, and exchanging quizzical looks with a fellow investigator. Then, she tells the teacher: "An onion makes your eyes cry." Thus, beginning with an *unspoiled attitude*, curiosity, and experimentation, Sheila is able to make a stark statement of fact and of distinguishing a category of crying—revealing to us a child's nearness to *the attitude of the scientific mind*. But while serving to enlighten the teacher, the incident does not suggest a formal science curriculum to be imposed by the teacher. Instead, it suggests further opportunities for child-initiated investigations, which motivate curiosity.

Watch three year old, serious-looking Matthew—clearly propelled by curiosity.

Lying on the nursery school floor, he intently watches a slowly moving earthworm, which he had brought in with earth in a paper cup. With furrowed brow, Matthew is studying the worm's movement. Abruptly, he turns to the teacher with a question: "Does a worm crawl?" The teacher answers "yes," assuming she is confirming the child's own observation. But her answer doesn't do. Matthew goes back to his own study, which soon leads to a conclusion: "A worm—he doesn't crawl; he s-t-r-e-t-c-h-e-s," and the precision of the word is sharpened by hand and arm gestures and eyes wide open, showing a stretching of the mind towards science.

What then are the conditions and the areas of school life and the immediate environment that naturally engage children's curiosity?

Weather

Weather is an area of daily encounters, practical challenges, discoveries, and pleasures—excitement and caution. In confronting weather, there is recognition of hazards in severe weather, conveying to the children measures we take for their protection. Unfortunately, many teachers are apt to be indifferent to weather as a handy source of significant learning and to prefer their own accustomed comfort of staying inside. Thus, they may prevent children from acquaintance with the **elements**, which are:

The wind. Ralph, age two and a half, is hardly experienced in ways of the wind, yet the word is in his vocabulary. The teacher, outdoors with the children, notices Ralph's awareness of the fresh breeze; she discerns his confronting it with his whole body, his breath and gestures. She sees no need to intrude. As the wind persists, the child is visibly moved—physically and emotionally—by the phenomenon and comes up with words for his absorbing experience:

"I like the wind. The wind hugs me," Ralph says (and pauses).

"The wind blows over my hair" (both hands on head and then some reflection and a circling motion of arms).

"The wind blows everywhere."

That was a brief episode with no lesson, but with self-discovery. The child was learning freely, in his own style—but relating to the teacher on the scene.

In another playground, Gregory, age three and a half, is prancing around, proud of his new blue shirt. Suddenly he notices that the shirt is moving while he is standing still. The teacher notices the puzzlement on Greg's face.

"Hey, Greg, what's the matter with your shirt?" she asks jokingly. But Greg is serious now, as he answers: "The wind's doing it." A fact he discovered himself. Like Ralph, Greg was getting acquainted with the wind in a physical, personal way.

When the teacher expresses her own curiosity, thus functioning as a cooperative learner while providing suitable materials, the children will be encouraged to be discerning and to be curious about various attributes of the wind.

Propulsion and direction can be felt and seen by simply holding a cloth like a sail, and moving with the force of the wind, which will lead *towards* understanding of sailing. Leaving different objects—paper, feather, stick, stone, etc.—on a surface outdoors or indoors by an open window in the presence of wind is another way of watching the wind at work.

The sun. The sun is yet another element of weather that children come upon in their daily life, as they experience its many attributes.

Photograph by Nancy P. Alexander

"Look!" cried Robert, pushing his plate aside. "There!" And he pointed to the floor in the middle of the room. "The sun fell on the floor!" His attention was caught by a bright patch of sunlight that had suddenly entered the room. Freddie, too, looked at it deliberately and

noticed the shadow of his head as he shook it and nodded. Harry got up to see, and there appeared his shadow cast by the sun on the floor. Several children squealed with surprise. A couple other children looked over the table on the clear sun-lit space and noticed the shadows of their moving hands—a new and amusing sight to the three year olds. But Robert was especially thrilled. He got on his knees in

front of the *fallen sun* and rubbed his hands over the area and made gestures of picking it up. "It's the sun!" he said emphatically.

Rain and snow. The physical presence and phenomena of rain and snow can invite concern and provide sensory (even sensational) activities that often arouse curiosity and stimulate thinking. Anticipating rain or snow and assessing weather is common adult talk which is familiar to children. But there isn't always an opportunity to have a personal encounter, to actually

handle rain or snow, to find out for one's self what those elements are.

Watching a mild, warm rain, the children were curious and anxious to get out—and they forgot the admonition to "watch out for wet spots." The teacher rushed ahead to the slide, sponge in hand, but it was too late. The first eager slider plopped right into rain water at the curvy bottom.

Subsequently, the children themselves established a temporary routine of delegating one child every morning to check on the outdoor equipment "in case it rained at night." There was no need for the teacher to extract a lesson from the situation. The children had done that themselves. The lesson was for the teachers— children need more opportunities to become personally acquainted with rain.

When interest was evident, and rain was actually *pouring*, the teacher put an empty pail out to collect it. In a short while she brought it in— splashing full! And, except for the established health rule of not drinking rain water, each child was happy to have a separate container of the water to do things with—to *wash* chairs and hands, to *water* house plants, to *compare* rain with water from the tap, to *detect* impurities, to *wipe* the spills.

Snowfall is a most spectacular demonstration of change in weather. Watch the snow flakes, falling steadily in silence; behold an unrecognizable world when all familiar objects are nowhere to be seen! Properly attired, there is no end to children's exploration, to first-hand contact, to making a bodily impression by sitting or lying down (when the snow is dry), or making and identifying tracks. Contact by poking with a stick;

attempting difficult shoveling; scooping snow and filling receptacles; compressing it; and, of course, squeezing snow to make a (crumbling) ball. Yes, there is struggle and frustration in managing the snow, but the result of throwing (or only thrusting and dropping) your own snowball is worth it! Curiosity becomes *ardent* when it's personally motivated.

Teachers need not focus on language arts or look for science concepts. However, teachers must not overlook the *children's* absorption with their own exploration, their persistence in spite of difficulty, and the delight that they experience—for these have bearing on the scientific attitude that children cultivate over a period of time.

Water and ice. Two three year olds were carrying a small pail of water to the sand box and spilled it on the paved part of the play yard— perhaps by accident and perhaps thinking *let's see what will happen if the pail is tipped*. What happened was fascinating to them! Some laughed gleefully; some peered straight at the puddle of water which changed dimension and direction and *moved*. "The water is going some place," said one. "It's going farther and farther," said another. "It wiggles." In no time, several children produced cans of water, poured it on the pavement, and proceeded intently with their study of the movement of water. Some children experimented with various obstacles which they introduced to "stop the water"—a board, sticks, their own feet— getting some notion how they could affect the movement of the water. They continued pouring water and watching it *go*.

One frosty autumn morning, children noticed the familiar water pail in the yard with something

Photograph by Francis Wardle

Plants Are Different

A child once asked me, after sniffing a sprig of mint, "Can you *eat* mint?" Not having a "yes" or "no" answer, I fumbled for words. But three year old Bella, who apparently had given as much thought to the question as I had, answered emphatically (perhaps even instructively): "You do not *eat* mint. You only *chew the smell out of it*." How precise a young child can be after sensory experience, and when challenged to think and speak!

Since any plant has many complicated parts and its life many complicated processes, young children cannot be expected to absorb more than one feature at a time—as they come upon it personally. A celery stalk can be cut into juicy chunks with scissors, but a coconut takes many whacks with a hammer or a rock before a crack exposes its milky juice. Roses have (ouch!) invisible thorns, along with (oh!) pretty flowers. The bark of the pine tree has a strong scent and a sticky pitch. The dry brown potato is moist and white on the inside. Fresh peas have a different number of peas in each pod, and no two peas are alike (contrary to a popular saying)!

Having any plants nearby, or at hand inside a classroom, the teacher can see how children's misconceptions are formed, and try to correct them.

A child was telling a friend where acorns come from, explaining the location on the ground. "I know

other than water in it. "Ice!" came an excited announcement from a four year old, followed by close inspection.

"It's not water any more."
"It doesn't shake."
"It doesn't spill."
And, after manual inspection, "It breaks."
After more poking and getting hands wet, a question: "Will the ice turn to water again?"
"If it gets warmer in a sunny spot," the teacher answered.

The children took it to mean a promise. When it actually warmed up and the ice changed to its former state, the children were as responsive to the teacher keeping her promise as to the evidence of water's characteristics. Although adults, including teachers, would take presence of ice in late fall for granted—nothing to get excited about—it's a different matter for preschoolers. This was a group of alert, articulate suburban four to five year olds who apparently had

previously handled ice only in the form of cubes in a drink, and even then, the ice was *served* to them, as part of the drink. They had not previously *handled* ice in the raw and in their own place.

Following the initial *finding* of the ice, the children wanted to "make *real* ice cubes" but weren't sure that would be possible without the Frigidaire. "Real" apparently meant the kind produced (by itself) in an ice tray. With the teacher's supervision, the children filled familiar ice trays with tap water and left them, covered, in the yard. To their surprise and jubilation, there was solid ice in the trays—without electric refrigeration—worthy of a proud report to parents!

The spontaneous project, including ice melting and refreezing, commanded the children's attention on and off and was remembered the rest of the term. Children spontaneously come upon many other characteristics of water in work or play.

that's where they are," he said convincingly. Fortunately, it was easy for the teacher in that school to arrange a short trip with a few children. As the teacher had surmised, the children had simply not *seen* the acorns *on* the oak tree or falling *from* it, but only lying on the ground. When some acorns actually fell with a thump, the children's response was: "Wow!" It was an exciting excursion. When the knowledge gained was shared, it became intensified, and was remembered.

Stones for Handling, Discoveries, Creativity

My own favorite natural item that attracts and holds my attention are stones—stones calling forth in me an instinctive response to their infinite variety of shape and surface, their universal presence, their abundance on beaches everywhere in the world; an aesthetically rich material that induces reflection; provides pleasure—all to be had, to be picked up, free! Moreover, my own appreciation of stones prompts me to think that such an *unstructured* raw material would elicit curiosity and invite safe investigation and various uses by children of a wide age range. What is required of the teacher is to have an experimental and open mind as to what might happen when children put stones to use in the classroom.

Obviously, the teacher's interest in a particular material will affect the children's interest and constructive use. When I placed a strong wooden box with approximately 20 stones of different shapes and sizes on the table, the box was immediately noticed by the children, and several came up to inspect the stones, remove them from the box, and ask where I'd found them. They definitely did not take them to be "educational play things." Practically all the children expressed some curiosity about the stones and spontaneously commented on the physical attributes—weight and shape—and compared the stones with familiar objects—such as a potato, an egg, a pancake, etc. The sharpness of one stone was investigated. Ideas for using the stones came from the children over a period of a few weeks. They served as cargo for toy dump trucks and pretend food items in the housekeeping area—especially for placing small stones in a pot with water to cook. There it didn't take long for the children to learn how stones take on a shiny look when wet.

In a few weeks of trying out the use of stones as a classroom material,

Photograph by Nancy P. Alexander

the teachers and the children could see how best to regulate the care of this particular material and assess the potential for children's learning in a preschool setting.

Since many children enjoy making some decoration on a finished block building, stones were used effectively for that purpose, sometimes with just one special stone. So we placed a box of stones on the block shelves. A tin can with stones was supplied in the *kitchen*. Some heavier kinds of stones were kept in the supply shed outside and used to hammer in an emergency or as a load when operating a pulley. Once a child deliberately buried a stone in the sand box and then pretended to have *found* it! The process seemed to be a simple game, yet it reflected a child's notion of what discovery is.

Summing Up

The following self-initiated activity of four year old Darius is particularly revealing of what John Dewey called "ardent curiosity" in children.

After watching a gardener digging out shrubbery roots, Darius went to the back yard, found a sharp stone, and proceeded to "dig for roots." "Look, here is a root!" he said excitedly, holding a stringy grass root in one hand and the stone shovel in another. He became so preoccupied doing it that he was reluctant to leave for lunch and talked about "different kinds" of roots.

Later, coming home from the beach, Darius immediately went into the back yard to resume digging, but this time he chose to use a shovel. He kept digging laboriously, until he succeeded in making a sizable hole and another discovery. He rushed into the house: "I found a lot of ants in my hole!" He continued watching the ants with absorption and amusement for the rest of the afternoon.

In the house, the subject of roots was still on his mind.

"What are roots for?"

"To hold the plant in the ground. Roots are strong."

"Yes! Roots *are* strong. I had to pull and dig hard to get them out."

After supper Darius went to digging again, and again the first thing in the morning. Impressed and satisfied with the depth and dimension of a hole, he asked for water to make mud. How good the mud seemed to him! With it, his fervor cooled; he relaxed, his intellectual-physical passion seemed spent—in pursuit of tangible roots.

And in his concentration in digging for roots, for revealed space, and for crawling ants, he dug a path towards science. He also knew how to relax from his intense efforts, the way many children do. And the way scientists must as well.

(Rudolph and Cohen. **Kindergarten and Early Schooling.** Englewood Cliffs, NJ: Prentice-Hall, 1984, p. 174.)

References

Dewey, John. **How We Think.** Boston: D. C. Heath, 1919, Preface.

Rudolph, Marguerita. "Scientific Elements Apparent in Spontaneous Activities of Nursery School Children—A Teacher's Account." M.S. thesis, Bank Street College of Education, 1954.

Marguerita Rudolph has a wealth of experience working with and learning from young children in and out of school. Being a devoted writer for more than 40 years, she has created many books for and about children. One of the books, Should the Children Know?, has been widely reviewed and translated in several languages.

Helping Children Ask Good Questions

by George E. Forman

Helping children to ask good questions is a matter of helping children to think about their thinking. We should help children be curious about their own understanding of events. But how do we help children look at their own thinking, since thinking is such a private event?

Consider this question: "Why does it rain?" The question is ill defined. Is the child asking for the purpose of rain, meaning, "What does the rain do?" Or is the child asking about the relation between clouds, moisture, and rain, meaning, "Where does rain water come from?" And, of course, it is possible that the child is only trying to engage the teacher socially. In our confusion, we are tempted to follow the child's question with another question, such as, "Do you mean why do we *need* the rain?" And we have all watched the child leave the scene on that one.

So how do we help the child think about his/her thinking without sounding too technical? We can ask the child to draw the event. Drawings allow us to look with the child at the child's understanding. Here are some real examples from the preschools in Reggio Emilia, Northern Italy (see note at end of article). Children from four to six years old were asked to walk through the city streets as it rained. They took photos of the rain and made audio recordings of the sound of the rain falling on the ground, the metal hood of a car, and a child splashing through puddles. Upon their return to school, they were asked to draw their understanding of the rain: its source, where it goes as and after it falls, how rain is used. They drew curves and spots to represent the sounds of the rain. At first glance, one might think that the children were being asked to give answers. But as you will see, the broader objective can be to help them think about their understanding.

Look at this drawing by Matano, age five and a half years. He has

Figure 1. Matano's rain machines

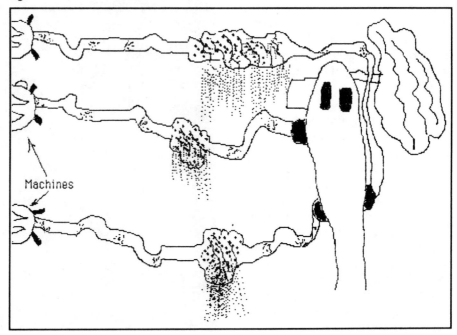

drawn a set of pipes that circulate the water from cloud to cloud. He explains: "In the sky there are some kinds of machines that never end . . . if one gets emptied there's another one ready right away. They have sort of pipes, the pipe goes into the middle of the clouds; there's water inside the clouds, the water inside the clouds never ends, if it ends it's no good, because then it wouldn't rain anymore. Then the pipe full of water goes back into the machines, where there's a sort of lever and when the machine is full of the water that the pipes brought, it switches on and it starts raining."

Matano, with obvious diligence, has represented his understanding of the rain. It is not important whether Matano believes that pipes are actually in the sky. What does matter is that Matano is trying to understand how water moves around in the sky. His best guess is that the water is forced through some conduit that is itself a rigid support. The idea that water might move airborne, without a rigid conduit, most likely violates his well held belief that water is heavy and would fall without support. A teacher who sees this drawing can be impressed with Matano's inventiveness within his own belief system and also acknowledge the incomplete nature of his understanding.

He has a budding notion of an eco-cycle, "some kinds of machines that never end," and of circulation of water. But his eco-cycle is an infinite linear chain rather than a cycle of reconstituted water; and his construction of circulation omits the lighter than air properties of water vapor. How do we help Matano ask the questions that will move him toward a more complete understanding of rain? To answer this, we need to make a shift in what we believe a question to be.

The child's answer is a question. That is, an answer that comes from the child's reflection on a previous answer is, in most cases, a type of rhetorical guess. The teacher follows up on Matano's drawing—first general praise, than a specific question: "This is interesting. Tell me why we need these pipes in the sky." The child might say: "So the rain can get from cloud to cloud. The clouds have to get rain from some place." Teacher: "And where do the pipes get the rain?" Child: "From the machines." Teacher: "And the machines?" Child: "That's a good one." We have used Matano's own construction of the event to address his incomplete knowledge and hopefully have helped him re-question his answers. The teacher's dialogue also serves as a model for a cognitive strategy that the child will internalize as he/she matures as a thinker.

One might ask, reasonably so, what is the difference in asking children,

"Tell me why we need these pipes" in the above example versus, "Do you mean why do we need the rain?" when the child had asked, "Why does it rain?" At one level, both teachers are asking the child to reflect on his/her thinking. The difference is in the directness. The question about the pipes allows the child to remain within the system that the child is trying to understand. The question about the question makes the child think about the difference between an intended message and the message received by the listener.

The latter carries an implied failure to communicate. In the example of Matano, the analog would be a teacher asking, "Do you really think that there are pipes in the sky?" The good intention is to have the child think about what he really thinks, but the effect will be for the child to think about what the teacher thinks of how he thinks. The teacher has caused the child to leave the system

Photograph by Subjects and Predicates

of events that the child is trying to understand. Thus, it is better to ask, "Tell me why you need the pipes." This takes for granted that the pipes are there for a reason, albeit that reason needs to be given to the teacher. The child's construction is honored at the same time that it is questioned.

The teacher tries to stay within the system of events that the child wants to understand. Thus, the way to help children ask good questions is to help children reflect on the facts until the child discovers something new to question. In this manner, the question that the child creates will emerge from his/her own construction of the system. The teacher is not literally teaching the child to ask the teacher better questions, but to ask better questions of the set of facts. And, since these sets of facts are the representations held by the child, the child is actually asking him/herself better questions. The dialogue indeed sounds like the child is giving answers; but as the child relates the answers to each other, a new question will emerge from the child him/herself. We can surmise that Matano now questions his facts— Where does the machine get its water?—and realizes that he has a system that contains a bit of magic, i.e. something comes from nothing. He could have easily accepted the magic by dismissing the teacher's last question with the conclusion: "The machines make the rain," thereby obviating the confrontation with magical thinking.

There are additional reasons why we want children to represent as much of the facts as they can—say through drawings, an extended description, or any other symbolic medium. It helps the teacher see the full context of the child's naive theories. The teacher is aided greatly by the richness of an ex-

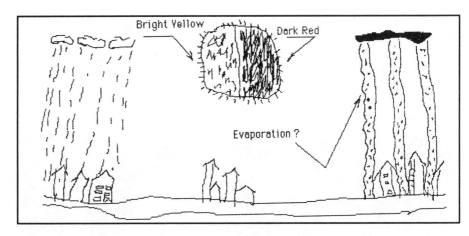

Figure 2. Simone's rain cycle

tended representation by the child him/herself. Let's take a look at another child's representation of the rain.

Here is what Simone (age five and a half) says about the rain as represented in the above drawing:

"The rain falls on the houses, on the umbrellas and on the trees, then it goes on the earth, in the courtyards. Then things dry up when it's sunny. Instead of continuing to go around the houses, the rain dries up, it heats the rain that has fallen and that's how it goes away afterwards, it goes back in the clouds and then it starts to rain again."

Simone's verbal explanation sounds complete. Unlike Matano's linear pipes, Simone's eco-system is circular. The rain falls, it (the sun) heats the rain that has fallen, the rain goes back in the clouds, then it starts to rain again. But since the teacher had asked her to draw this system, as well as describe it, we notice some interesting gaps in her understanding. Notice that the return of the rain (right side of drawing) portrays the rain flowing up inside bounded channels. Might we have here another version of Matano's pipes based on an incomplete understanding of airborne vapor? Does Simone's "dries up"

simply mean "moves up"? Does "the rain goes up" mean rain as liquid?

The observant teacher has several opportunities to help Simone ask better questions about her construction of the rain cycle. "Tell me about these heavy lines here (pointing to the thick lines bounding the flow of rain back up into the clouds)." Or the teacher could say: "I see that you have drawn the rain between the buildings here (on the right) but all around the buildings here (on the left). Tell me about that." Or: "I see that your sun is half bright and half dark. What's happening here?"

On this last question, the teacher wonders if Simone will comment on why evaporation is happening on the dark side of the sun. There is also the possibility that Simone has confused a whole earth symbol for night and day with a more suitable representation for sunny and shade in a given city. This is a subtle confusion between the shadow of the earth itself versus the shadow of clouds in a city. We should not expect too much progress on this distinction at age five.

Nevertheless, the teacher can help Simone begin to consider the differences between night and shade. A more complete understanding will dawn later.

In each case, the teacher tries to understand the child's understanding and then gingerly asks the child to tell more about an aspect of the drawing. The observant teacher will chose an aspect of the drawing that has the potential to lead the child to discover gaps in understanding.

When done with skill, this teaching technique keeps the children interested in their own struggle to understand. At the same time, the teacher allows the children to experience the personal satisfaction that comes when they first discover the gap and subsequently when they reconstruct their understanding of the event.

Note: These examples were taken from the exhibit "The Hundred Languages of Children" that is touring the United States in San Francisco, Fort Worth, Syracuse, Amherst/Westfield, and Boston during 1988-89. The drawings and comments by the children are real examples; the teacher follow-up questions are suggestions by the author.

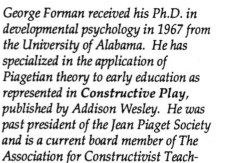

Photograph by Nancy P. Alexander

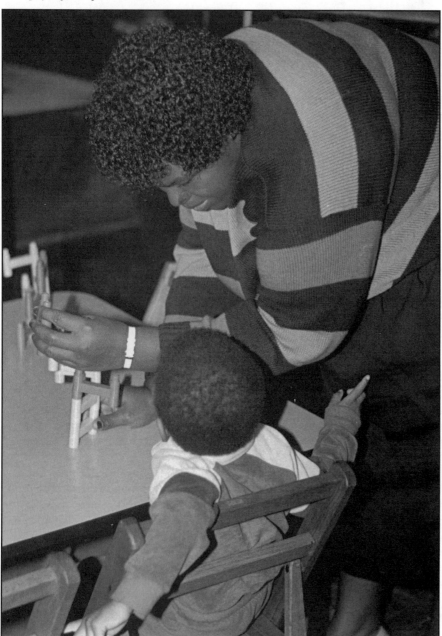

George Forman received his Ph.D. in developmental psychology in 1967 from the University of Alabama. He has specialized in the application of Piagetian theory to early education as represented in Constructive Play, published by Addison Wesley. He was past president of the Jean Piaget Society and is a current board member of The Association for Constructivist Teaching. His current work deals with the use of graphic representation to help children understand their own thinking.

Teach Science? Me?

by Nancy P. Alexander

You **can** teach science. In fact, you are most likely teaching science now, even if you do not realize it. Many activities in a good early childhood program are science but may not be recognized as such. Teachers frequently feel their own scientific knowledge is inadequate and that they cannot therefore convey accurate information to children. Yet, teaching science is teaching thinking skills; the very nature of science makes it a vital component of any curriculum plan.

Teaching strategies used to help children learn through exploration and discovery are cultivating the methods of the scientist. Providing opportunities for children to identify problems, make predictions, experiment to seek solutions, and analyze their success is a matter of using materials in a way that promotes children's cognitive skills. These are the methods of science; the teacher need not feel insecure.

1. Set Goals

Before any learning center is incorporated into a preschool program, the teacher must determine what the center will be designed to accomplish, since the goals will define the manner in which the learning center is organized and arranged. Science centers can be open-ended, offering free exploration of materials and their properties, or they can be designed to encourage the child to discover a specific phenomenon.

One water activity may consist of simply a large pan of water with various items for pouring and filling; another might be designed to help children conclude whether certain items will sink or float. Regardless of the purpose of a center, "Please, do touch" should be the message conveyed. "Look, listen, smell, and, when appropriate, taste—use all your senses" should be the message received by the child. "Explore, experiment, and enjoy."

2. Become Articulate

Teachers should identify the concepts children can potentially learn from an activity or center and plan how they can lead children toward understanding those concepts. Becoming familiar with the appropriate vocabulary will enable teachers to use terms accurately. This preparation will make it possible to respond and react to the interests of the children, and to follow their lead as an activity evolves.

3. Use Resources

The juvenile, non-fiction 500 and 600 sections of the library are a good resource for information and activities. Since these books are intended for elementary-age readers, they are an easy-to-use medium for compiling information and often include activities and experiments which can be used (or adapted for use) with preschool children. The many books of science activities currently available for early childhood teachers are also good resources for specific experiences (see resources chapter—page 84).

The teacher's challenge is more one of selecting from the myriad of possibilities than of locating information. A general topic such as water, for example, can generate dozens of variations of free exploration of water, and just as many experiences designed to develop a specific concept related to water properties.

Photograph by Nancy P. Alexander

see, and make inferences noted. The teacher must be alert to the sometimes subtle signs that children are losing interest, then determine whether the activity should be varied, expanded, or ended.

Case in Point

Establishing science centers is essentially setting the stage for exploration and discovery by providing the necessary materials and supplies, arranged attractively and made accessible for the children's independent use. The essence of a science program is an environment designed for hands-on learning, supportive of the compelling curiosity of young children, and accompanied by appropriate guidance from the teacher. The following episode illustrates how this can be accomplished:

When a group of four year olds arrived one morning, they found a science table with a few pennies in a basket, two small bowls—one with water and one with vinegar—and some small pieces of cloth. The teacher observed silently from a distance as a group gathered around the table. The children recognized the coins as pennies and several noted Abraham Lincoln's profile. They wondered about the clear liquids, and one child commented on the "pickle" smell. "It's vinegar," another contributed. "My mother uses it on lettuce."

Soon, as the teacher had anticipated, a child put a penny in the water. In time, another was placed in the vinegar; the children watched excitedly, intrigued by the chemical change taking place on the coin in the vinegar. "It's getting new!" a child commented as they observed the chemical reaction transpiring. Just as the teacher had envisioned, all the pennies were soon dropped in the vinegar and "shined." Two

4. Invite Participation

Effective science areas require an arrangement of materials that attracts children's interest, helps them recognize possibilities for use, and inspires them to experiment. Even outdoors, an array of new items for molding, a special container for fresh dirt, and an attractive pitcher of water will invite children to an area designated for making mudpies, while they will avoid a muddy, jumbled pile of

equipment left from yesterday's mudpies.

5. Stay Alert

The teacher must be constantly alert to how children are functioning within the center. Their success in using materials, making observations, formulating provocative questions, and creating ways to test their ideas should be carefully observed; their abilities to see relationships, describe what they children soon tired of the activity and left to pursue other interests, but as other children gravitated toward the table, the teacher was prepared with more pennies in her pocket to add to the supply. Some children were content to watch without participating. Several delighted in telling others that the vinegar would make pennies shiny, but the water would not. Two busied themselves with seeing how many pennies they could shine, counting them, and forming neat stacks with the finished products.

Each child participated at a level of involvement appropriate for that individual's interest. The teacher had provided the preparation, support, and sanction for the children to learn on their own— allowing them to question, experiment, and draw conclusions themselves. The actions of the children had been accurately predicted. It was not necessary to tell them what to do nor how to do it. The innate curiosity of children was allowed freedom of expression, the desire to explore was uninterrupted, and the teacher had *taught* science in the most effective way possible. We can pre-stage such motivational events as this, but many unplanned occasions for science discoveries also occur regularly in our daily life with children.

The Teacher's Role

What qualities does the teacher need to provide proper guidance for children's science experiences? An understanding of how children learn and a sincere desire to provide them the opportunities to learn in the way they learn best are two prerequisites. An inquisitive nature and the ability to enjoy *with* children the spirit of adventure and sense of discovery that are an integral part of the process of science are valuable assets. There must be an under-

standing of the teacher's role as a guide rather than as an imparter of knowledge, as a facilitator of thinking rather than a conductor of information, as a supporter instead of a controller.

The good early childhood teacher creates an atmosphere conducive to the development of each child's self-esteem through a positive, accepting environment where curiosity and the testing of ideas are actively encouraged. Listening with undivided attention to a child's question or explanation, supporting the child's captivating desire to experiment by providing the necessary materials, sanctions, and freedom from ridicule will encourage science experimentation. The teacher's own interest, energy, and enthusiasm are indispensable—more important in a quality science program than the most expensive equipment or the most elaborate plans.

A Teacher in Action

How does a teacher serve as a guide for young children's learning? The following vignette illustrates how one teacher guided a

child's fascination with disappearing snow:

"The snow's gone; it's all gone!" was the disappointed observation made by a three year old who had experienced snow for the first time. The teacher could have reacted by merely stating, "It melted; it turned to water." Instead, the teacher knew of a shady portion of the playground where snow remained, as yet unaffected by the rise in temperature, and suggested the two of them go to see it. Allowing the child time to notice the wetness from the melting snow and giving permission to bring a bowl of snow inside to observe, the teacher then queried, "What do you think would happen if we also put some in the freezer where it is cold like the weather was yesterday?"

For the next several hours, the three year old periodically checked the bowl of snow placed on a tray in the classroom and the other bowl in the freezer, eagerly sharing with friends the exciting news that the snow in the freezer did not turn to water, but the snow on the table did. At playtime, the children were allowed

Photograph by Subjects and Predicates

to wear rubber boots and play in the wet slush still remaining outdoors.

The child had a question, not stated as such, but a wonder recognized by this astute teacher. The teacher, resisting the temptation to offer a simple explanation, instead helped the child to find out—essentially carry out an experiment in a manner understandable even at such a young age. The teacher did not give answers, but led the child to find answers by providing a responsive environment and guidance in making associations.

Finding the Words

As this teacher did, we can help children develop observation skills through their innate, driving curiosity about the world around them. We can respect the wonder and awe they have for their environment and help them express that wonder with language.

The ability of young children to learn the language to which they are exposed is impressive, and science experiences often are the incentive and motivation for learning new words. We should encourage children to verbalize their discoveries and supply them with the necessary vocabulary. Young children can understand and use such multi-syllabic, adult-sounding words as *dehydrated* and *evaporation* as easily as they can learn *dried* when the words relate to their own rich experiences.

The vocabulary of science should not be introduced as any type of lesson or directed learning but used deliberately by the teacher in appropriate context where the children can interpret meanings and associate words with the concepts being explored. Language should be a friendly social accompaniment to activities, not a formal lesson or memorization of terms.

The teacher's role, then, is to establish the climate conducive to exploration and discovery, to guide that exploration and assist in the communication of discoveries. Within this role is the power to sustain curiosity by allowing it to flourish or to suppress it irretrievably. In this role is the power to increase critical thinking skills, including the ability to make predictions, judge, and analyze—to give children the world or keep it from them.

Nancy P. Alexander is executive director of Child Care Services, Inc. of Northwest Louisiana. She received a master's degree from Northwestern State University. She is active in professional organizations, a regular contributor of photographs and articles for educational publications, and has given numerous presentations at national and regional conferences.

Supporting the Development of a Scientific Mind in Infants and Toddlers

by Judith Leipzig

Alana, a ten month old, is sitting in her high chair. There are some pieces of banana on her tray and a bottle of juice, and her caregiver is feeding her some baby cereal. With a very serious look on her face, Alana begins to squeeze several pieces of banana in her fist, allowing them to squish out between her fingers. She smears the residue all over the tray. In between squeezes, she pops a few pieces of banana into her mouth. As the caregiver proceeds to feed Alana the cereal, the baby leans over the side of her high chair and drops some of the remaining pieces of banana on the floor, peering intently after them. She then grabs her bottle, pounds it on the tray, and throws it after the banana pieces.

Does this anecdote fit in with the image you have of a *scientist*? As adults, most of us probably equate science with test tubes and Bunsen burners, with telescopes and microscopes, and at the very least with the sophisticated use of computers. We have mixed up the tools a scientist has and the data she must have at her fingertips with the

workings of her mind. Technique and high tech materials have gotten confused with our understanding of the mind of the scientist. It is as if we said that expensive kitchenware and a pile of cookbooks make a master chef, while the truth lies in recognizing that without special sensitivities to taste, color, and texture, to the variations of palate and combinations of materials, no one could create really fine food.

Yet, this is what we're doing when we think that in order to start children off on the right track in this high tech world, we need to work on producing *superbabies* from the first day of life. Flash cards, infant stimulation programs, toddler learning centers, and so-called "scientifically designed" toys all hype the extreme benefits of themselves. Teachers and parents are being sold a bill of goods that focuses solely on the collection of miscellaneous information and the use of tools, without any regard for the basic workings of the mind.

If we examine what children are working on cognitively in the first three years of life, we recognize that there are really two ways of looking

at learning. One could say quite simply that a child learns, for example, how to open a cabinet door, or how to count to the number ten, or the names of things; and all this is true. But the real underlying strengths and skills that children are working on encompass these relatively straightforward pieces of information.

In the first three years, babies and toddlers are learning to learn. They are learning ways of collecting information, integrating it, and making use of it. They are learning to ask questions, and to analyze answers; and all this is part of developing a scientific mind.

The younger the learner, the less we can separate the cognitive from the emotional, the physical, the social—all these are intertwined. The very qualities that make for a gifted scientist are really, at heart, the qualities that describe a creative artist. The mind that reaches out to take in the world, to turn it over, to make sense of it, and to describe it in a very unique way is the mind of any truly creative person—whether we describe a molecular biologist, a painter, a psychotherapist, or a teacher.

This article will examine some of the qualities of the scientific mind, and the kinds of interactions between adults and very young children which support the development of such a mind. There are two themes in the understanding of human growth and development that run throughout the exploration of these interactions. First, children need to be grounded in strong relationships, and they need to feel safe and connected in order to function at their highest cognitive level. It is not the best techniques, the most elaborate equipment, or the most effective curriculum design that is the strongest foundation for children learning. The real facilitator of a young child's growth is the tuned-in presence of a trusted adult. Research tells us that children who are with a trusted person are better able to explore the environment, make new relationships with others, and take pleasure in the world. Children need to have the home base of a relationship from which to explore the rest of the world.

Second, young children learn by their own actions in the world—by the actual experiences they have doing, feeling, mastering. Piaget and many others tell us that infants and toddlers learn through their senses and through their bodies by moving, mouthing, banging, stroking, carrying, exploring. The more actively a child can be involved in a situation, the more chance she has of gaining a true understanding.

Following are some of the qualities teachers need to support to foster the development of a scientific mind.

The Drive to Explore and a Sense That It's Worth It to Continue Learning

One of the gifts of infancy with which we are all born is a tremen-

Photograph by Bonnie Neugebauer

dous drive to take in and understand the world. Babies and toddlers do this for its own sake, often apart from any other idea of a goal. Michael Lewis and Susan Goldberg beautifully describe the special pleasure a baby has when she makes something happen, when she feels what they call "the joy of being the cause." This sense of oneself as a competent person, and as a person who has excitement about learning and growing, comes in large part as the result of *contingent responses*—that is, responses specifically as an answer to a child's own gesture.

Contingent responses by adults are acknowledgments of the child's communication, or answers, or when possible and appropriate, the accommodation of the child's interest in active engagement. This requires that adults are actively listening, so that they don't miss the message. For example, a common contingent response to a baby's crying would be coming to pick the child up, or feeding him, or changing his diaper.

A contingent response does not mean that we give children everything they ask for. Say, for example, that a two year old is demanding to fool around with the arm of the stereo turntable. Do we damage this child for life, turning him into an apathetic fellow, if we deny him the opportunity to scrape the needle across the record for as long as he likes?

The answer, obviously, is that we can't give children what is either unsafe or inappropriate just because they ask for it. But the contingent response here might be, "I see that you're very interested in the record player, but it's very delicate and it's not for children. Let's put it away and find something else for you. How about scraping this spatula across the xylophone?" What the teacher is saying is: "I hear you. I see what you're interested in, and I value that interest even if I can't allow you to continue the behavior." The toddler may express some disappointment and frustration, but will not feel invisible as a child might feel who was totally denied.

Another kind of contingent response that a caregiver could provide for a child might be something like, "I see you want the nesting blocks, but I can't get them down for you this minute. As soon as I'm done diapering Jennie, I'll be there to give them to you." Again, the teacher values the child's interests while she reasonably adjusts her response to the needs of the entire group.

Babies need to feel that what they do matters, that their gesture or communication causes a response of some sort. This is why busy boxes are so interesting: push a button and you've made a bell ring. And this is why people, who with our facial expressions, our voices, and our actions are the ultimate busy boxes, are so important and powerful from the first moment of life. By responding to babies quickly and consistently, we're teaching them that it's worth it to communicate.

The Ability to Concentrate

There is a special kind of focus that scientists bring to bear upon the subject of their investigations. This is an intense beam of interest, an absorption which cannot be deterred by the enticements of less important activities. We have an image of the serious scientist working in her lab into the wee hours of the morning, absorbed in her work in a way that makes her disregard all else. This ability to concentrate on a matter of interest is, in part, something that many babies are born with. Babies are paradoxical—on the one hand, much of their learning is global and undifferentiated, but, on the other hand, it is an incredibly minute examination of experience. How many adults would not only notice, but also explore, the variations of feeling when brushing up against blanket edge and blanket middle? It

is the rare adult who would display the impressive ability to become absorbed in a question that the toddler shows when he pushes a chair back and forth from one side of the room to the other, while he learns about space and movement and himself and more.

Adults can either assist children in developing their concentration further, or we can teach them to derail their trains of thought in midjourney in much the same way that television commercials interrupt our absorbed experience of a story and program our minds to expect these breaks from cognitive engagement.

In this area, one of the important things teachers can think about is learning the skill of watching and waiting. It's important for adults to recognize when **not** to interrupt as a baby crawls in and out of a box on the floor repeatedly, or a toddler concentrates for a full five minutes on sponging off an already clean table. Children need time to process their experience, so if they continue to look absorbed, we can know that they are still working on learning and integrating information. They also need the time to gain a sense of completion in the task. When possible, a teacher should think about whether that baby really needs her diaper changed this very minute, if it's absolutely necessary to go to the store right now, or if the adult can honor the child's absorption for a little while longer.

The Ability to See Relationships and to Think Symbolically

Scientific discoveries are made, in part, by a rich knowledge of the components of a situation, and the ability to see the relationship between one phenomenon and another. Doctors who noticed that certain medical and behavior

problems were most displayed in children who lived in particular kinds of housing led to the discovery of the connection between lead poisoning and peeling lead-based paints. We would not have penicillin today if Alexander Fleming had not noticed that a mold which was growing by chance in his petri dish was prohibiting the growth of his original topic of study, staphylococci bacteria.

There are a number of ways that teachers can support the development of the skill of seeing relationships. Babies and toddlers need indepth experiences with many different aspects of the world. This does not mean that we cram in classes in music appreciation for the two year old, or read political science texts to babies inutero. On the contrary, it means that we give children extended opportunities to explore those aspects of their world that will have the most everyday meaning for them. Children need the chance to do things over and over again: to eat snacks, and listen to the wind in the trees, and put their hands under running water. They will absorb information in a deep way if they can do so at their own pace, at their own level, following their own interests. What we want to do is set up an environment that beckons to children rather than commands them. Materials which can be used for symbolic play can be made available—both things which are open-ended, like play dough, paint, and water, and objects which can be used to role play about the real world.

Extended sensory experiences are necessary but not sufficient. In order to be able to recognize relationships, we have to have a knowledge of the various pieces, and the ability to step back from our immediate experience of them to see the bigger picture. Teachers may find that they "naturally" sometimes ask the questions that Irving Sigel calls "distancing questions" and which he suggests promote the development of symbolic thought. These are questions that stress the non-present, the abstract, and the numerous facets of any object or experience. They give the child the chance to make sense of her world by participating actively, by predicting, discussing, recalling the past.

For example, a 17 month old and her teacher are out for a walk. Suddenly, a dog races across the street in front of the stroller. The teacher comments, "Oh, look at that doggy! Doesn't he look like your dog Frisky?" The child is presented with a question which requires her to reach back into her mind, pull the image of her dog Frisky from her memory, and place it symbolically in front of her right next to the street dog. The question pushes her to compare and contrast the two. It is not necessary that this toddler be capable of actually completing the task or, even if she can, of answering the teacher. The important thing here is that she gets some experience in stretching a cognitive muscle, and building up her capacity to use her perceptual experience in an organized and meaningful way.

As teachers pose distancing questions, they have to be sure that they pause to allow some space and time for the toddler's mind to reach for an answer. A 30 month old at the water table begins to play a game with his teacher. Holding the rubber duck above her head, the teacher calls out, "Heeeere comes the duck! What's going to happen when it falls into the water?" She pauses a moment before she lets the duck splash into the water. In that moment, the toddler is being called upon to anticipate what will happen in the future. Again, the correct

answer, whether it's articulated or not, is not the goal. Rather, the teacher is giving the child the opportunity to briefly step back from his sensory experience and to exercise his newly forming skill.

Another kind of distancing question asks the child to reach for options and alternative ways of looking at something. "Is there another way we can put this hat on? Where else could you put that hat?" the teacher asks a 23 month old who is sitting on the floor fooling around with an old cowboy hat. A game might ensue, with the teacher and other children suggesting increasingly silly ways to wear the hat: on your foot, on your belly button, inside out.

One point to stress: all of these distancing questions have to be asked in the course of relaxed and sometimes playful conversations, in the context of the child's interest. If the questions are actually tests or drills, they lose their effectiveness. When pressured to perform, children will stop functioning at their highest level and will retreat from engagement.

Flexibility of Thought and the Courage to Investigate New Ideas

An important aspect of creative work is the ability to approach an old problem from a new point of view, or to ask a question that has never been asked before. Some major scientific discoveries have come about as the result of a scientific mind taking the same pieces and fitting the puzzle together in a new way. The scientist has to have a certain courage, a belief in his capacity to search for and possibly find a better answer. Copernicus used the same components that others had used before him: the planets, the sun, and the stars. But

his flexibility of thought permitted him to imagine a different way of perceiving the workings of the universe. He was able to play with the accepted tenets and to develop the idea that the planets revolve around the sun, not around the earth. It took a great deal of courage for him to go intellectually where no one else had gone before and more courage to discuss his thoughts with others.

Adults can contribute to this flexibility of thought and courage to investigate by encouraging children's unique and creative ways of learning about the world. Children need to be in an accepting environment, one which cherishes individuality. If an adult demands that an infant or toddler adhere too strictly to both rules of social conduct and to rigidly defined ways of thinking, then we end up with limited thinkers. The adult who abruptly takes the hairbrush out of a baby's hand as the infant taps a toy piano with it, in order to teach her that hairbrushes are for one thing only, or the teacher who makes fun of the child who mixes his juice with his green beans is pulling the rug out from under the development of a scientific mind.

Our acceptance and approval of the individuality of children fuels their courage. If kids experience investigation as tricky or rigid, or learn from us that there is only one correct way to do things, we will stunt their intellectual growth. When a toddler wears a shoe on his hand and a teacher mocks him or sharply corrects him, he is being sent a message of limitations. Toddlers know where shoes are really worn. What they are doing, when they use them to carry tiny dolls, when they pretend to drink out of them, when they throw them like a ball, is experimenting with the possibilities, investigating the nature of things. In open-ended play situations, there are as many right answers as there are individual thoughts. Infants and toddlers need to get a lot of experience looking at things from different viewpoints, making their own discoveries, and feeling good about themselves as competent explorers, while they are collecting information about the ways of our world.

A Sense of How to Approach Problems

Children can get more mileage out of some kinds of learning than they can out of others. The toddler who learns to recite a series of numbers is highly unlikely to really have an understanding of the concepts of quantity and numbering. The baby or toddler we referred to in the beginning of this article who learns how to open one cabinet door or recite a list of French words does have a piece of information. In contrast to these kinds of learnings, consider the person who has learned how to approach a problem, how to compare that question with other similar and dissimilar situations, and how to go forward in solving the problem. This person will be able to use the problem-solving skills to learn about many other things, and to gather and integrate concepts in new and possibly unexpected situations.

Teachers can actually teach problem solving skills. What we want to do is to begin to lead children through our thinking, and break down for them the steps of our problem solving processes in situations that have real meaning for them. To do this, we have to be alert to the opportunities for sharing this skill. For example, in the middle of a busy day in a child care center, Sara, a two year old, is pulling a toy dog on a string. Suddenly she begins to whine energetically. The teacher

notices that the string is caught around the leg of a chair. If the teacher swoops in, unwinds the string, and lets Sara go merrily on her way, then the child may learn something about the teacher's availability to her, but she will not have learned anything about problem solving—except perhaps that a whine sometimes brings a solution.

But let's imagine that the teacher has enough time and staff support to take this situation as a chance to begin to develop in Sara some problem solving skills. Instead of giving answers, or doing *for* the child, the teacher needs to look for ways of including the child in the process. She may ask, "What's the problem here?" When she asks this question, she puts a kind of frame on the situation. Often, for very young children, distress has a kind of global feel to it: "I was pulling the toy along the floor and everything was good, and now everything is bad." Sara may not be able

to actually identify at this point the source of her unhappiness (that the dog can't be pulled) or even spot the fact that the string is wrapped around the chair leg. We begin to illuminate for children the process within our own minds by identifying that there's a specific cause to the distress—in this case, that the toy is stuck; and this is the beginning of teaching an approach to problems. The teacher says, "Let's figure out how to do this together," thus telling Sara that she's there for her, that people can work together, and that Sara's concerns and interests are legitimate.

Sometimes talking through problem solving with toddlers is almost like talking out loud to oneself. On the one hand, the language and the concepts do have to be tailored to the developmental stage of the child; but on the other hand, the teacher asks questions that the child may not actually have the answer to. In fact, the point here is not only to teach a toddler about disentan-

Photograph by Francis Wardle

gling strings from chair legs, but more importantly how to go about identifying, articulating, and researching solutions for questions. The window to her own mind that the teacher has opened in order to let Sara watch the mechanics of problem solving might go something like this: "Oh, I see there's a problem here." (pause) "What happened to the toy dog?" (pause) "Hmmm, it looks like he won't go. It looks like he's stuck." (pause) "What is he stuck on?" (pause) "Oh, his string is wrapped around the bottom of the chair. How shall we get him out?" (pause) "How shall we unwrap that string?" (pause) "I think I'll get closer so I can take a good look. Sara, can you help me pull this string out? I'll pick up the chair and you pull the string."

In this kind of interaction, the teacher needs to remember several things. It's important to pause briefly after each thought, a kind of contemplative, considering pause. Like the distancing questions discussed earlier in this article, these comments or questions need to be followed by an opportunity for Sara's mind to reach for an answer. If the teacher gives Sara the chance to be actively involved, both with her mind and her body, this experience will have greater impact on Sara. Second, as in the distancing questions, we ask these questions not as a test, but as a way of sharing, in a chatty way, how we go about the business of approaching this task. Third, Sara needs to experience this kind of teaching repeatedly; and eventually she may be able to answer some of the questions you pose, before you provide the next step.

It is the accumulation of the experience that deepens its positive effect on Sara—no child under the age of

three will learn the problem solving approach simply by one clear explanation. At a later point, Sara may begin to be able to identify the problem—"Stuck!"—and still need help with the other steps in the solution. No matter what her level of participation is, whether it is simply listening while you think out loud, or saying "Stuck!," or finally getting to the point of knowing what to do about a tangled string, Sara needs the teacher to actively and repeatedly compliment her on her thinking. "You did a good job. That was good thinking." Or "We really worked on that one, didn't we? I liked the way you used your words to tell me about the problem." Sara needs the teacher to underscore many times during the course of her first three years how much thinking and grappling with a question is valued. And last, if a child is in real distress, the teacher, clearly, would not proceed with this kind of technique. No one can pay attention to learning a new skill when tired or very upset.

It's easier for most people to be able to imagine a "Science Institute for Babies," a program where adults try to structure all of the interactions and scale down preschool and even elementary school curricula, than it is to envision the potent work of the developmentally tuned-in teacher. To the untrained eye, much of the work of a really good teacher has a kind of invisible quality. It is so subtle, and flows so much from the teacher's reading of the individual child's abilities, interests, and needs, that it may not seem active enough to the person who thinks that teaching means filling up a vessel with data.

This article has discussed some of the ways that important adults can support those qualities of a child's mind that will contribute to grow-

ing abilities to be competent discoverers. By valuing the experience of an individual child, and by helping children to see the meaning in their efforts to connect with and understand the world, adults are providing infants and toddlers with the most important foundation of all. Cognitive development is so interwoven with every other domain that, in order to foster the qualities of the scientific mind, we have to nurture the flowering of the whole child.

References

Lewis, Michael, and Susan Goldberg. "Perceptual-Cognitive Development in Infancy: A Generalized Expectancy Model As a Function of the Mother-Infant Interaction," **Merrill-Palmer Quarterly**, 1969, 15.

Piaget, Jean. **The Origins of Intelligence in Children**. New York: International Universities Press, Inc., 1952.

Sigel, Irving E., and Rodney R. Cocking. **Cognitive Development from Childhood to Adolescence: A Constructivist Perspective**. New York: Holt, Rinehart and Winston, 1977.

White, R. W. "Motivation Reconsidered: The Concept of Competence," **Early Childhood Play**, edited by M. Almy, Selected Academic Readings. New York: Simon & Schuster, 1968.

Judith Leipzig is a specialist in infant and parent development. Currently, she teaches at Bank Street College of Education in New York City.

Meaning Through Process: Science for Children with Special Needs

by Patricia Scallan

Observe a typical three year old engaged in an endless pursuit of earthworms in a wet sandbox. There is no doubt that the nature of a young child's interactions with the world is one of exploration, discovery, and manipulation.

In classrooms where the atmosphere is trusting and nurturing, young children are drawn to the magic and mysteries of our physical environment. Teachers in touch with the processes of a young child's thinking can gently guide the discoveries of the young scientist, asking questions, sharpening observation skills, and extending ideas. Through this process, the child learns to connect experience and information leading to the development of concepts and the construction of knowledge.

For early childhood educators, one valuable goal is to begin to teach the elegant process of scientific thought—to develop young children's abilities to pose difficult questions, seek answers, analyze information, and arrive at meaningful conclusions. Simply stated, in teaching science we give kids a chance to think and to contemplate what has been experienced, to label

it in their own words, and to make the experience a personal memory upon which future impressions and knowledge may be drawn.

The opportunities and benefits derived from early childhood science curriculum are no less significant for teachers of children with special needs—children with physical, mental, or emotional handicaps; development delays; specific learning disabilities; as well as the unidentified and extensive population of children who are educationally at risk.

Special Benefits

Through science, children with special needs are exposed to wonder, knowledge, and appreciation of nature and the physical world. This experiential learning can become an important foundation for understanding and communicating basic information, which will, in turn, strengthen the connections and interactions between children with special needs and their normal peers.

Thinking skills. In fact, science activities provide a context for the

child with special needs to develop essential thinking skills—such as observing, describing, and classifying—and problem solving skills—including formulating questions, inferring, and concluding—that will later be applied in other areas of curriculum. When we use science content to develop and reinforce the basic learning and functional skills of children with special needs, we increase both their intrinsic motivation and the relevance of what they are learning.

Social context. Children in therapeutic settings often lack basic knowledge upon which human interactions rely. Play skills are often inhibited or nonexistent for these children who have not had the broad exposure to social interactions common to play groups of normal children. Because impaired children may be developmentally behind as much as 18 months or more in crucial areas like expressive or receptive language, processing and organizational skills, motor development, and social interactions, they frequently have not developed the same knowledge base of their peers. It is from the shared knowledge of peers that rich and varied play interactions can evolve. Messing about in science can provide the context for special needs children to enter the world of their peers.

Multi-sensory experiences. Many children with special needs require a multi-sensory approach to learning. Early childhood science curricula create the perfect blend of direct exploration through hands-on learning, repetition, feedback, structure, and language skills. For a substantial number of children with special needs or at-risk children, the capacity to use language skills to organize or construct meaning is seriously impeded or impaired. When carefully planned, science activities create an essential framework for developing key vocabulary, for organizing information and sensory input, for communicating ideas beyond the spoken word, and for practicing and applying skills presented in therapeutic settings.

Individual needs. The essential nature of science exploration also provides the flexibility teachers need to adjust to the range of levels and abilities of their students with special needs. The hands-on nature of preschool science invites the constant manipulation of concrete materials, further reinforcing the variety of learning modalities—auditory, visual, and sensory motor—that are critical to taking in and processing information.

Safety and comfort. When science learning is structured as an individual or small group activity, the performance anxiety of the child is diminished. Confidence and self-esteem are reinforced. Greater opportunities for participation exist and the fear of failure is lessened. The inherent structure of science skills and concepts easily lends itself to modifying or breaking down of activities into smaller, sequential, more manageable tasks.

Photograph by Nancy P. Alexander

Teacher Issues

Although the movement toward mainstreaming children with special needs is gaining momentum, regular classroom teachers are hesitant to teach the handicapped or developmentally delayed child. Classroom teachers are expected to make reasonable accommodations to these children while at the same time promoting appropriate integration of regular classroom curriculum. In many school systems and centers, the process of working out specific curricula and adapting activities is left to the teachers. Consequently, teachers react to the challenge of mainstreaming with trepidation: How much extra time is involved? What will I have to do differently? How will other chil-

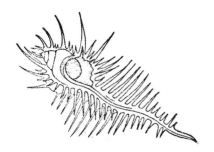

dren react to the pace and needs of a special child?

Alan Gartner and Dorothy Hergner Lipsy, in "Beyond Special Education: Toward a Quality System for All Students" (**Harvard Education Review**), conclude that one major impediment to mainstreaming is the lack of supportive resources, such as modified curricula. They contend that when modified curricula is available, teacher confidence and effectiveness is increased.

The task of adapting curricula need not be formidable when teachers recognize that the positive outcomes of teaching science in the classroom are the same for the child with special needs as for the child who learns through regular channels. The goal of science for all children is to open and expand the avenues of discovery for each child.

Structuring for Success

Special needs children particularly benefit from the self-paced, discovery approach that is rich in sensory input and concrete activities. Teacher interaction is structured carefully to develop specific and basic process skills such as classifying, using number, measuring, using space-time relationships, communicating, predicting, and inferring.

Different paces. The most important aspect of teaching science to children with special needs is to structure each activity and outcome so that children meet with success.

Central to this idea is an awareness that the pace will differ. Children with special needs require more time to explore and manipulate materials. They need to handle the objects before observations can be made and connections perceived. For example, it may take several days of dropping plastic toys in the water table before the impaired child recognizes that not all objects sink to the bottom of the water table.

Small units. Children with special needs require that the tasks be broken down into smaller units. An activity with magnets involving classifying objects on the basis of attract or repel would need to be approached as two separate sets of activities. Several activities would center around observing and classifying items that are attracted to magnets. At a later time, activities focusing on objects that magnets repel would be presented. Subsequently, the teacher might experi-

ment with putting the two concepts—attract and repel—together, and plan activities that require the child to classify using the two concepts.

Many opportunities. Children with special needs will require more opportunities to gradually interact with the materials and participate in activities. Teachers can be open to the possibility that discoveries can be reinforced through the day-to-day experiences such as creating green from the accidental mix of yellow and blue paint at the easel. Interest centers that invite casual discussion and observations throughout the day become essential time for special children to interact and absorb information.

Clear definition. Because organization skills are less refined in many children with handicaps or potential learning disabilities, teachers need to assist each child in defining what he sees. Vocabulary should be

Photograph by Michael Siluk

specific and clear. Tasks need to be presented in careful and simple sequence. All items that will be used in an activity should be assembled and organized in advance. Teaching objectives should be clear to the child as well as the teacher.

Concrete and visual support. Children with difficulties in language and processing skills can be assisted through the use of visual supports and concrete objects. In recalling what sinks and what floats, the teacher can facilitate memory and organization skills by having the children glue the actual objects tested on a classification chart. Photographs of what sank in the water table will help children with auditory or visual memory problems to access and recall information.

It is important to remember that children with special needs may not be able to make inferences as quickly as a normal child. They need to see the whole process before offering an idea or hypothesis. In cooking projects, if the child can observe first hand the chemical changes through a glass container, he will more easily comprehend the process and descriptive comments of the teacher and peers. Often teachers rely too heavily upon oral instructions and observations, not realizing that these children fail to comprehend the discussion. Visual support and aids are strong reinforcers to verbal information given.

Repetition. Children with special needs will require more repetition, in concrete ways, using a variety of sensory channels. They may need to act out, several times, the caterpillar feeding on leaves, spinning a chrysalis, and emerging as a butterfly before fully embracing the concepts of transformation and metamorphosis.

Teachers can facilitate learning by structuring output in a variety of ways with different modalities. Some children cannot communicate their results using words nor can they answer a direct question, such as, "What happened to the very hungry caterpillar?"

Opportunities to build models with blocks and clay; to draw with paint and markers; to indicate an answer by pointing to pictures, concrete objects, or photographs; to recreate meaning through creative movement, mime, and play acting are all valid means of communicating knowledge than can extend and solidify experiences for young children. When children are unable to speak what they know, they may be able to mimic and dramatize it.

Focusing on the Process

For children with special needs, the process of science exploration is as important as the end result. Spontaneous observations and direct exploration may be sufficient to provide the context for applying specific skills and knowledge these children have acquired outside the mainstream class.

Children with special needs must address the *what* questions as much as the *why* and *how*. Science has the intrinsic flexibility to allow children to comprehend materials at their own pace and in their own way. It sustains their attention and, when carefully structured, can assist each child in developing self-control, initiating ideas, organizing and executing a plan, and communicating results.

Each part of the science process—whether it be observing, classifying, or manipulating materials; acquiring new vocabulary; or interacting with other children—is an important learning tool for children with special needs. To provide a context whereby children can ponder nature's mysteries, and can label and express discovery in their own way, all the while personalizing and internalizing that experience—this is the true essence of science and learning.

Patricia Scallan is director of the National Child Research Center in Washington, DC.

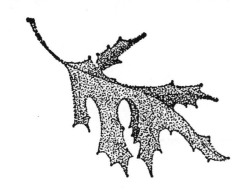

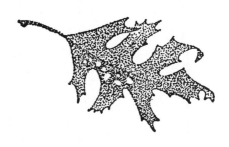

Seeing Possibilities
Everywhere—

Ideas for Science
Learning

My Fingers
by Mary O'Neill

My fingers are antennae.
Whatever they touch:
Bud, rose, apple,
Cellophane, crutch—
They race the feel
Into my brain,
Plant it there and
Begin again.
This is how I knew
Hot from cold
Before I was even
Two years old.
This is how I can tell,
Though years away,
That elephant hide
Feels leathery grey.
My brain never loses
A touch I bring:
Frail of an eggshell,
Pull of a string,
Beat of a pulse
That tells me life
Thumps in a person
But not in a knife.
Signs that say:
"Please do not touch,"
Disappoint me
Very much.

Setting Up an Enticing Science Interest Center

by Karen Miller

In a good early childhood program, science happens in many places, not just in a designated science interest center. Swinging on the swings and riding tricycles are exercises in physics. Cooking involves chemical changes. Children notice the weather and what the sun does. Paint evaporates. Science learning opportunities are available in many of the everyday activities and spontaneous happenings in the classroom. However, a special place set aside to help children grow in awareness of nature and other science concepts can be the focal point of your classroom—a place for exciting discovery and endless investigation. Here are some questions to ask of your science corner:

Is it enticing?

Is it an attractive display? Do the materials in the center seem to *invite* children's participation? Think about how department stores entice people to examine goods by attractive, eye-appealing displays. You could take the opportunity to play up some of the aesthetics of nature.

Children could help you make a mobile from natural objects gathered on a field trip.

Instead of having shells, rocks, and pine cones in a heap, they could be sorted into attractive baskets, or children could be challenged to line them up according to size.

Large tree stumps make attractive display tables.

A solid block of styrofoam is good for displaying feathers. The end of the feather can be stuck into the styrofoam, displaying the feather in a vertical position.

You can make a display screen by taping together three or four panels of corrugated cardboard, such as from grocery store boxes. These could be painted or covered with fabric or contact paper. Materials and pictures can be attached to the screen, and it can be stood up on a table top. In addition to making an attractive display, this screen can help serve as a divider from the neighboring interest center.

Is it well organized?

Is it clear where things go and what belongs there? Can children look for and find materials they want to use? Can they help themselves?

It is ideal to have a large table for display and work space for children, as well as a shelf for storage of materials. If you don't have enough shelf storage space (is there *ever* enough?), multi-compartment shoe boxes from the closet department of a variety store make good display and storage spaces for small objects. You could even use the compartmentalized boxes available at liquor stores, spruced up with contact paper, for an interesting display.

Clear plastic bins or shoe and sweater boxes make good storage boxes for your science area. Children can see what's inside.

You can develop *kits* for various science activities, and children could help themselves as their interest dictates. This works especially well for kindergartners and school-agers. All the materials for bubbles could be in one box. Magnifying glasses and numerous materials glued to index cards to view with the glasses could be in another; food coloring, eye droppers, and ice cube trays for color dilution experiments in another. A list of materials that belong in the box could be attached

Photograph by Francis Wardle

to the box, and children would notify the teacher when certain supplies are running low.

Is there something to do?

In a *typical* science center, we might see some sleeping gerbils or hamsters buried in cedar shavings, and perhaps an aquarium with some goldfish in it. On the table might be some dusty pine cones, some shells, a few rocks, a balance scale, a large magnifying glass, and maybe a color paddle. There are only things to look at, but nothing to hold the child there more than a minute or two. Each interest center in the classroom should have some involving activity to hold the child there for a little while. Otherwise, the center will not absorb its *share* of children, and other interest centers might become overcrowded. The problem can be easily remedied.

Give the children some specific challenges using the balance scale. First of all, let them play with the scale without directions, and then ask them some questions about it. "What makes this side go down?" "Why doesn't it go down when I place this rock on the other side?" When they seem to understand how it works, you could pick one object to be used as a *standard*, an apple, for instance. Let the children guess which objects in a random collection of materials on the science table are heavier than the apple and which are lighter, and put them into two piles in front of the scale. Then let them test their guesses. Children must first use their senses for this activity. To keep the activity interesting, you could select a different object each day to be the standard. Lining things up from lightest to heaviest is a more complex activity, because children have

to compare objects to each other rather than just to one specific object. It takes considerable flexibility in thinking and is quite difficult for pre-operational children.

Sorting objects is one easily available *thing to do* at the science table. Science involves observation, noticing details, and categorizing materials—all activities children practice when they sort things.

There are two types of sorting. One type is when the teacher selects the categories: "Put all the rough stones here, and all the smooth stones there." This type of sorting is a type of *review* or *test*. The teacher can see if the child really understands the concepts; and the child might learn the concepts, especially if another child who does understand the categories is participating at the same time.

The other type of sorting is where the child selects the categories. All the teacher has to say (if anything) is, "Put the things together that go together." Providing a muffin tin, a compartmentalized tray, or boxes into which to sort materials will encourage interest in the activity. Then the teacher can ask the child, "Why did you put these things together?" and let the child generate the language to describe the categories. An extension would be for the teacher to ask, "Can you think of a different way to sort these things now?" Again, this requires flexible thinking on the part of the child.

Put two bowls of water on the science table. Add ice cubes to one. Let the children transfer the ice cubes from one bowl to the other using kitchen tongs. They will enjoy the manipulative activity and will see that the ice cubes get smaller as they melt at room temperature.

There are, of course, many more activities to involve children in science concepts. The resource chapter in this book will lead you to many more ideas.

Is there something new?

If it's just the same old stuff day after day, it will fail to hold children's interest. At least once a week, something should be added or changed. As well as alleviating boredom, this encourages children to ask questions and helps develop their curiosity.

Try adding several different types of flashlights one week. Perhaps you could cover the table with a blanket and children could go underneath with the flashlights. They will enjoy the mechanical aspects of the flashlight. The next week you could add colored transparent plastic (book report covers are great) and let children discover a new effect.

A *nature show and tell* is an excellent alternative to *bring and brag*. Children can bring in items from nature and show them to the other children during your group time. On the spot, let the child dictate a *museum* description for the object. "This is Jason's Rock. He found it in a stream where he was wading with his mother. It has sparkles and it turns a darker color when it gets wet." These objects are then added to the science table, along with their labels, that children can read to each other or classroom visitors.

Are animals and plants teaching what you want them to teach?

Many early childhood classrooms have pets—guinea pigs, rabbits, gerbils, hamsters, snakes, chameleons, birds, fish, hermit crabs, etc. It is good to frequently remind yourself of why we have these animals in the classroom. They can teach children about responsibility and what it takes to keep something alive, content, and healthy. If the teacher does all the maintenance of the animals—cleaning the cage, feeding, etc.—the children are losing much of the learning experience. Worse, if the animals are neglected, cages are dirty, or children are allowed to harass and poke at the animals, the lessons being learned are very negative. By all means, involve the children in the regular care of the animals and talk about why you are doing things.

It is natural for children to want to touch and pet the animals. This gives you the opportunity to talk about gentleness, even empathy. Tell children that the animal thinks they are *giants*. This, again, requires flexible thinking for people who have always considered themselves the smallest type of person. It allows them to treat animals the way they would like to be treated.

Let children create an *experience chart* about their classroom pets.

Display the chart near the animal and read it frequently with the children. Children will enjoy reading the chart themselves for parents and other classroom visitors. One group of children saved all the feathers they collected from the bottom of the cage of a molting parakeet and made a chart showing the different types.

Likewise, neglected plants teach children the wrong lessons. Generally plants are not terribly interesting to children. It takes too long for things to happen. Help children notice changes in their plants. If they select the young plant themselves on a field trip to a greenhouse, or plant seeds, their interest will be higher. Count the leaves. Make charts. Have watering schedules to check off.

A bulletin board that grows with the plant is one good idea when you plant beans. Start with brown paper to represent soil and a white *seed* shape in the brown. Have the children observe their beans every day and tell you what is new. As the seed sends out shoots and the

Photograph by Nancy P. Alexander

first leaves form, continue to add these representations onto the bulletin board, with the children participating. Eventually the bulletin board will show a mature plant with bean pods hanging from it, ready to harvest.

An amaryllis, available in winter months, is a wonderful plant to grow in the classroom. It starts with a large bulb that you plant so that half the bulb is visible above the soil. It quickly sends up one or two shoots and grows as much as an inch a day! Children can measure the stalk every day. After about a month the end of the tall stalk opens into spectacular trumpet flowers. Children will be dragging their parents over to see it.

Am I interested in science?

You can tell what a teacher values by what she has in the classroom—how much space is allowed for it, the attention paid to the presentation. As with any type of early learning experience, the involvement and interest of the teacher will have great impact on children's learning. You can provide the *envelope of language* around children's sensory experiences. Most important, be a good model yourself. If you place a high value on the science corner and exhibit genuine curiosity and wonder at natural objects and phenomena, children will exhibit these same attitudes.

*Karen Miller is early childhood education editor for Scholastic Inc. and a contributing editor to **Pre-K Today** magazine.*

*She is author of three books, **The Outside Play and Learning Book**, Gryphon House Inc.; and **Things to Do with Toddlers and Twos** and **Ages and Stages**, Telshare Publishing.*

Giving Children What They Need to Investigate Electricity
A Supply List from Larry Morris

carbon batteries, size D (buy at least 4)
battery holders (that will hold size D batteries)
alligator clips
knife switches (buy 1 or 2, DPDT)
light bulbs (not flashlight), 6 volt (buy 2 or 3)—threaded E-10 lamps
   or bayonet style
bulb holders (socket base for threaded E-10 lamps or for bayonet
   lamps)—they must match bulb type
buzzer, 6 volt (dc)
solderless spade tongue terminals
solderless ring tongue terminals
wire, 22 gauge
crimp tool—for fastening terminals on wire (could also be vise grips)

"Sciencing— It's Here, There, and Everywhere!"

by Karen Stephens

Begin sciencing with young children—with an emphasis on the "ing"! That little participle transfers a static curriculum into an exciting process of learning through discovery and active investigation. When working with young children, sciencing adventures can be found everywhere and anywhere you turn. Children are positively driven to understand themselves and the world in which they live; it's their work and they embark on it with a vengeance. Children pose questions relating to science all through the day. They are intensely interested in where things come from, how things are made, and how they can put them to use. Many sciencing projects can spring forth from the children's motivated inquisitiveness. They not only want to know, but *need* to know about their world and how they fit into it.

This article presents many diverse topics for sciencing inside and outside the classroom. It is designed to help you "brainstorm" many more ways of guiding children in the process of discovering not only themselves; but also their world *and* their ability to appreciate, understand, protect, and enhance it. Happy sciencing!

Air Around Us

- Discuss why humans, plants, and animals need air. What is in air that we need?

- Put a sponge into water and squeeze. Note the air bubbles that surface. Did the sponge float before the air was squeezed out of it? Did it float after the air bubbles surfaced?

- Have some children kneel on a deflated inner tube; pump it full of air. What happens?

- Play with streamers, kites, pinwheels, and whirli-gigs on a windy day. Wind is moving air.

- Provide three jars ranging from small to large. Put a candle in each. Light the candles and chart how long it takes for each to burn out. Why did this happen?

- Watch clouds moving by. What makes them move?

- Blow up balloons, then let the air rush out onto your hands.

- Watch smoke coming from chimneys. What makes it swirl and sway?

- Stand in the wind. Then sneak behind a tree or wall. What happened to the wind?

- In outer space and underwater, there isn't air for people to breathe. What do astronauts and scuba divers do so they can still explore the outer and underworld? (Note: SCUBA stands for self-contained underwater breathing apparatus.)

- Have the children ride a trike with a flat tire. Fix the tire. Which way was the easiest to ride?

- Make hand fans to cool each other off on hot summer days.

Birds

- Discuss how birds are alike (lay eggs, have beaks and feathers).

- Discuss how they are different from each other.

- Listen to the sounds of birds.

- Discuss where birds live. Find one nest and visit it often. What happens?

- Discuss how birds feed their young and teach them to fly.

- Learn how birds help humans (beauty, sounds, eat insects).

- Investigate all types of nests (hanging, cliffs, cactus, etc.).

- Compare many bird eggs by size and color.

- Make bird nests from grass, mud, sticks, and string.

- Provide neighborhood birds with a feeder, bath, or house.

- Collect some bird feathers. Put a drop of water on them. What happens? Look at the feathers with a magnifying glass. Note the shaft and barbs.

- Hatch eggs. (Some local hatchers loan out equipment.)

- Display a bird skeleton or a picture of one. Discuss the hollow bones which allow them to be so light.

- Make a nesting bag for local birds. Fill an onion bag (netted) with string, tissue, yarn, etc. Keep an eye for birds taking the materials to build a home.

- Visit a bird or pet shop.

- Play a bird call record during nap. (The "Environments" records are best.)

Block Area Activities

- Provide or build: ramps, wheels, pulleys, levers, pendulums, etc.

- Highlight the importance of balance and weight distribution for stability.

- Discuss the force of gravity when a block building tumbles to the ground rather than flies up to the ceiling.

Noting Continuums and Cycles

- Mealworms to beetles (ask the local pet store)

- Egg to chicken

- Hospital birthing nursery to nursing home

- Seed to full grown plant bearing new seeds

- Cocoon to butterfly

- Sapling to giant sequoia

- Sheering sheep to woven fabric

- Tadpole to frog

- Sunrise to sunset

- Spring to winter

- Baby teeth to adult teeth

- Crawling to running and leaping

- Freezing to thawing

- Evaporation from lakes to rain from the clouds

- Explore momentum with ramps, balls, marbles.

- Provide a heavy box. Have the children try lifting it first. Then use a plank over a brick as a lever to help out with the work. Is pushing the box up a ramp easier than carrying it up stairs? Can a pulley make a bucket of toys easier to lift?

- Arrange several planks at varying angles for the children to walk up. Which is the easiest? Hardest? Is it as hard to go down as to go up?

- Have children roll a cube and a ball down a ramp. Which is first to the bottom? Why?

- Have children try to carry sand toys to the sand area. Then provide a wagon. Which requires the most human energy?

Ecology

- Make a pollution gauge by greasing a saucer with Vaseline and leaving it outside over a weekend. Check its appearance and contents when returning Monday morning.

- Provide two socks. Put one over the end of a car's exhaust pipe for two to five minutes. Compare the two. The sock has caught some of the material that goes into the air, but there is much more invisible pollution. (*Don't* bring the sock inside the school, the fumes are smelly!)

- Fill jars with water. Into each put one of the following: leftover food, tissue paper, soap detergent, car oil. Note what happens to the water. This is what goes into many of our streams and lakes.

- Go outside and just listen. What sounds feel good to your ears? Which ones don't?

- Collect bottles, tin, and paper and recycle them.

- Take a picture of a very littered area. Clean it up and take

---

Dramatic Play Kits for Sciencing Minds

Space Traveler
Veterinarian
Rock Collector
Jeweler
Doctor/Nurse
Florist/Gardener
Baker/Cook
Carpenter
Construction Worker
Chemist/Scientist
Scuba Diver
Robot
Plumber
Railroad Worker
Pilot
Fire Fighter
Farmer/Ranch Hand
Fisher
TV or Radio Announcer
Telephone Repair Person
Camper/Nature Hike Leader
Weather Announcer
Photographer
Bird Watcher
Red Cross Disaster Relief Helper
Forest Ranger
Piano Tuner
Bee Keeper
Mountain Climber
Auto Mechanic
Insect Collector
Vision and Hearing Screening
    Specialist
Small Appliance Repair Person
    (watch, clock)
Astronomer
Safety Inspector

---

another picture. Which is best for the environment?

- Go for a litter walk.

- Find ways of conserving water and electricity in the school or home—turn lights off, don't leave the record player running when not in use.

- Fix a broken toy in the center rather than throw it out. Let the children help.

- Sort through a wastebasket at the end of the day and talk about what could have been recycled for use in the art area or sand and water table.

- Help children learn about conserving fuel by helping their parents organize car pools. Perhaps walk to a park rather than ride a bus.

Energy

- Learn about all kinds of power: solar, wind, electric, gas, nuclear, and *human*.

- Practice conserving energy, i.e. walk rather than ride to a park, help children and parents arrange car pools, turn lights off in unused rooms, have lunch by candlelight.

- Compare battery operated toys to child operated toys. Have the children chart the number of toys in each category that they find at home or school.

- Compare bikes with automobiles, motor boats vs. canoes, snow skies vs. snowmobiles, hang gliders vs. jet airplanes.

- Recognize the sun's energy and its impact on humans, animals, plants, weather.

- Identify sources of human energy: food, exercise.

- Have the children conduct a survey of all the items in the center requiring an energy source other than the human hand, i.e. lights, furnace, electric typewriter, electric clock, dishwasher, electric can opener, electric hand mixer, aquarium pump and light, record player, tape recorder.

- Discuss the *energy* of rivers and streams which impacts the terrain, the energy of a storm that has passed through.

- As children swing, lift toys, dig in the sand, or run and play tag, their ability to use energy can be highlighted.

Gardening

- When tilling a plot of soil, note the different textures and objects, i.e. decayed leaves, pebbles, dirt, worms, bugs, etc.

- When thinning seedlings, follow the package directions for most of the garden, but leave a patch unthinned. Compare the results as the plants grow and mature.

- Plant natural pest controllers such as marigolds rather than use pesticides.

- If you fertilize your garden, leave part of it unfertilized and compare the results.

- Take polaroid pictures of your garden every few days. Let the children put them into proper sequence of growth.

- Plant some vegetables in a shady place as well as in a sunny area. Compare growth results.

- Vary the amounts of water given to different rows of plants. What are the results of too much water? Too little?

- Make a compost heap or mulch for natural fertilizer. (Scraps from snacks and lunch, leaves from fall raking.)

- Keep charts on the height of seedlings weekly. At the end of the season, make decisions *with*

the children regarding successful crops to be sown next year.

- Cut open your harvest—fruits, vegetables, flowers— and look at the insides. Talk about the parts of plants.

Field Trips for Sciencing Minds

Hot air balloon lift-off	Glass cutter's studio	Mountain climbing expedition
Kite flying contest	Orchard or tree nursery	Orchestra rehearsal
Crayon factory	Lumber yard	Laundromat
Scuba diving class	Telephone company	Car wash
Amish settlement	Bee farm	Toy factory
Caves (Spelunking)	Boat dock or wharf	Trolley or bus ride
Chemist's laboratory	Rock quarry	
Planetarium	Health food store	*. . . and don't forget . . .*
Greenhouse	Newspaper print shop	your own backyard
Recycling center	Meat locker	a nearby stream
Aviary	Gasoline station	a grassy hillside
Textile factory	Top of a *very* tall building	a star gazing camp out
Aquarium	Underground home	cloud watching
Rock and mineral museum	Bicycle race	a bird's nest
Pet shop	Poison control center	an ant hill
Florist shop	Nursing home	a favorite tree for all seasons
Natural habitat zoo	Baby nursery	a wild flower meadow
TV or radio station	Native American history museum	a buck eye tree in the fall
Bakery	African history museum	town square fountain
Recording studio	Eye doctor	sprinklers in the summer
Solar or wind energy display	Horse farm	butterflies around spring time
Whale watching site	Dairy farm	puddles
Animal humane society	Entomologist lab (studies insects)	walks in gentle rains
Veterinarian	Shoe factory	watching a thunderstorm roll in
Nature preserve	Robotics lab	a frozen pond
Artist's workshop	Computer center	earthworms after a heavy rain
Archaeological dig	Movie theater (tour *behind* the	a weeping willow tree in gusty
Sky diving class	scenes)	winds
Hang gliding launch	Restaurants	thick frost on bare tree branches
Space center	Hospital emergency room	crunching through paths carpeted
Welding facility	Fire department	with fall leaves
Carpentry class	Nautilus work-out room	watching squirrels run circles
Construction site	Dinosaur and fossil display	around tree trunks
Weather station	Dark room for photo developing	a spider's web with morning dew
Water-slide recreation center	Potter's studio	a rainbow tying a ribbon on a
Train or subway station	Chicken hatchery	summer's rain
Airport	Ice cream factory	an eclipse
Fireworks display	Automobile factory	the sunrise and sunset

- Save seeds from this year's harvest to use next year. Sort and classify the seeds.

- Sprout some of your seeds for salads, i.e. radish, bean, alfalfa.

- Preserve some of your harvest for later eating: freeze, dry, can.

- If you don't have garden space, plant tomato plants in large pots. Make a terrarium or an undersea garden in the aquarium. Desert gardens in large clay saucers show off cacti.

- If you plant a flower garden, pick some flowers to press and dry. Dried flowers can be made into arrangements. Pressed flowers can be added to homemade paper for unique stationery.

- When gardening, discuss the benefits of earthworms to break up the soil and fertilize.

- Try to plant as many varieties of vegetables as possible. Discuss the part of the plant that is usually eaten, i.e. root, stem, flower, leaf, seed.

- Talk about other creatures that like gardens (in particular, birds and rabbits). Problem solve means for protecting your crop— or be generous and share with the wildlife.

Mirrors and Reflections

- Provide many shiny objects to look into, such as a coffee pot, toaster, front and back of a spoon. Note the changes in the reflection. Do the children see a difference in their appearance?

- Use mirrors to look above, below, to the side, and in back of you.

- Use mirrors to reflect the sun. Can you make the reflection move around?

- Make reflector cards by covering 12 inch cardboard squares with aluminum foil. Take the cards outside and the children can play *reflector tag* by trying to tag children with their spot of light.

- Give children three mirrors to arrange in many different ways. How many images can they produce by moving the mirrors to different angles?

Rejects

- Take apart and explore the insides of an old clock, radio, camera, washing machine, hand mixer, or any appliance that is not usable.

- Unwrap an old baseball or golf ball.

- Find out what is inside a music box.

Sand Area Activities

- Build a *beaver* dam with gathered leaves, twigs, branches.

- Build a volcano. After the cone is made, insert a plastic tube into the top. Add vinegar and red food coloring, drop in teaspoons of baking soda and it erupts!

- Make moats and canals. Provide toys for traveling the routes.

Discuss why moats were built around castles. How do canals help with transportation of food?

- Hide gold painted rocks in the sand area so the children can go *panning for gold*. Talk about prospectors and miners. You can also bury geodes, fossils, etc.

- Hide *clean* chicken or turkey bones to be found and excavated by budding paleontologists. (Be sure to get them all so you don't attract animals.)

- When playing, talk about droughts and deserts.

- Build a city and then create a flood. What does the water do to the city?

- Provide sand and water wheels.

- Use the sand to build large sculptures. One group of children made a great dinosaur and used L'eggs containers for eyes.

- Hide metal objects in the sand and use a magnet or metal detector to find them.

- Make plaster casts from designs created in the sand.

- Talk about sand storms and their danger to visibility, eyes, ears, crops.

Shadows

- Ask the children when they can see their shadow. Go out in the sunshine and play with shadows. Can the children make their shadow big, small, wide, thin? Measure their shadows.

- Can you make your shadow shake hands with your friend's shadow without touching hands?

- Set a bottle in the sun. Draw around its shadow each hour.

- What shape of shadow do different objects make? Can you change them?

- Make a variety of *pictures* or images on a screen by putting objects in front of the slide projector light. Make silhouettes of the children.

Shoes

- Discuss the purpose of shoes: protection, health, sure footing, fashion.

- Collect many different shoes. Analyze what they are made of and why the specific material was used (wooden, rubber, suede, cloth, leather).

- Talk about the process of making shoes (leather from hide, measured, sewn).

- Label the parts of the shoe (heel, tongue, laces, toe, eyelet, sole, arch support).

- Discover different ways for keeping shoes on (laces, buckles, zipper, velcro).

- Practice caring for shoes (cleaning, airing out, buffing).

- Observe shoes from different cultures and discuss why they wear them (snowshoes, moccasins).

- Collect and observe recreation shoes and their purpose, i.e. bowling, tap, mountain climbing, aerobics, basketball, baseball.

- Put some water on many different shoes. Which ones repel water? Which ones would the children wear in the rain, in the snow, on the beach?

- Visit a shoe repair shop.

- Sort shoes according to color, size, use, type of material.

- Take a walk with one shoe on and one shoe off (around the play yard). How does the shoe help? What does it keep you from feeling?

Sounds Around Us

- Drop a penny onto different surfaces—metal, wood, water, fur. Which is the highest sound? Lowest?

- Strike a tuning fork and then put it under water.

- Listen to a clock through a curtain rod, yardstick, etc. What does it sound like?

- Put your ear to the floor and listen to sounds.

- Feel your throat while you talk and note the vibrations.

- What different sounds can you make with your body? Find other things that make sounds.

- Talk about the parts of the body that are used for speaking: vocal cords, diaphragm, tongue. Show pictures or diagrams.

- Use a tape recorder to record everyday sounds as well as the

children's voices. Play back for children to identify.

- Discuss sound pollution. Problem solve several means for reducing it.

- Provide instruments for children to explore the sounds produced with percussion, wind, and string instruments.

- Make wind chimes from found materials.

- Look at everyone's ears. Are they alike? Show diagrams of our inner ears. Highlight safety precautions to retain good hearing. Try to obtain a plastic model of the inner ear.

- Have children listen to each other's hearts through a stethoscope while at rest and then after running and jumping.

- Have everyone listen to each other's stomachs.

- Discuss warning sounds, i.e. siren, growling dog, thunder, rattlesnake.

- Play with echos in stairwells.

- Have a child close her eyes and then guess from where another child is calling her name.

- Hang different sized nails from a hanger and let the children *play* them.

Water Around Us

- Discuss uses of water: drinking, washing, cooking, extinguishing fires, playing in, transporting materials, feeding plants and animals, uses in art.

- Identify different bodies of water. Display a globe and two jars of water, one pure and one salt. Note the bodies of water on the globe. Which are salty? Which aren't? Let the children taste each sample and compare.

• Wet the back of the children's hands and then have them blow on them. Note the coolness. Discuss the purpose of perspiration.

• Fill a tin can with ice cubes on a warm day. Note the condensation.

• Note ways for humans to obtain water, i.e. faucet, juices, fruits and vegetables that have a high water count.

• Fill two jars to the same level with water. Mark the levels with a line or rubber bands. Cap one and leave the other open. Note the amount of evaporation as the days pass. Compare the jars.

• Mix salt and water in a pan. After evaporation, note the residue.

• Breathe on a cold window then note the water vapor.

• Fill a jar to the brim and cap. Freeze outside. Note that the jar cracks from the water expanding when frozen. (Remember safety on this one!)

• Drop water onto a paper towel, wax paper, aluminum foil, and different types of cloth. Which ones absorb, which ones don't? Which material would be best to make a raincoat?

• Make a rainbow with a hose. Note how the water droplets refract the light so its colors may be seen.

• Make a wave bottle. Mix colored water and oil in a *tightly* capped plastic bottle, such as a dish washing liquid bottle. When tilted side to side it makes waves. Discuss why this happens. Mix oil and water in a bowl to further develop the concept of *not mixing*. (You can put small plastic fish in the bottle for added interest. Cut the fish out of any plastic lid.)

• Make a *water magnifying lens*. Take a large plastic container (i.e., sherbet) and cut holes in the sides large enough to put a hand through. Loosely place plastic wrap over the top opening and secure with a rubber band. Carefully place water on top of the plastic wrap (it will sag a bit). Children place objects through the side holes and then look through the water lens to see them enlarged.

• Dip different fabrics into water (corduroy, silk, cotton, nylon) and then hang up to dry. Which dries first? Last? What happens to the water?

• Dip a ruler, pencil, and other long, thin objects into a glass of water. What happens?

• Compare the density of water to other liquids such as oil, honey, or karo syrup. Place each substance on a piece of wax paper. Tilt to an incline and see which one wins the race to the bottom.

• Make popcorn. The heated moisture inside makes it pop!

• Filter some muddy water through different fabrics. Save some of the muddy water and compare to the filtered water.

Topics to Nurture Sciencing Minds

Seasons (all together or separately)
Rocks, minerals, and geological formations
Planets and stars
Types of transportation
Types of communication
Insects
Reptiles
Jungle animals
Sea animals
Creatures underground
Nutrition and food
Colors
Air
Dinosaurs and fossils
Shells
Tools and machines
Weather
Human body
Natural resources
Birds
Sound
Water
Bodies of water
Trees and leaves
Flowers and plant life/seeds
Farms
Cities
Reflections
Volcanoes
Spiders
Ecology
Hospitals
Growth and aging
Sports and Recreation
Clothing
Maps
Pets
Pond life
Desert animals
Art
Types of occupations
Inventions
Types of homes: human and animal
Types of sports

Woodworking

- Classify types of wood according to name, hardness, or special uses, i.e. cedar to keep away moths.

- Provide woods of varying hardness for hammering. Which are softer? Harder?

- What is the effect of sandpaper on wood?

- Feel the heat of friction after sawing or sanding.

- Use a variety of tools for woodworking: saw, level, screwdriver, sander. (Keep safety in mind!)

- Explore ways of preserving wood: paint, tongue oil, varnish.

- Find out where lumber comes from.

- Count and learn about the rings on the hammering stump.

- Make log cabins out of dowels, build a maze for the pet hamster, make a roller coaster for marbles, make paddle boats, make a sign for the classroom door.

Bibliography

Austin Association for the Education of Young Children. **Ideas for Teaching with Nature.** Washington, DC: National Association for the Education of Young Children, 1973.

Blemming, Bonnie Mack, and Darlene Softley Hamilton. **Resources for Creative Teaching in Early Childhood Education.** New York: Harcourt Brace Jovanovich, 1977.

Busch, Phyllis. **The Urban Environment.** Chicago: Ferguson Publishing, 1975.

Carson, Rachel. **The Sense of Wonder.** New York: Harper and Row, 1965.

Day, Barbara. **Open Learning in Early Childhood.** New York: Macmillan Publishing, 1975.

Hall, Mary Yates. **Simple Science Experiences.** Canville, IL: Instructor Publication, Inc., 1968.

Harlan, Jean. **Science Experiences for the Early Childhood Years.** Columbus, OH: Charles E. Merrill, 1984.

Hill, Dorothy. **Mud, Sand and Water.** Washington, DC: National Association for the Education of Young Children, 1977.

Holt, Bess-Gene. **Science with Young Children.** Washington, DC: National Association for the Education of Young Children, 1977.

Lorton, Marjorie. **Workjobs.** Menlo Park, Ca: Addison-Wesley, 1972.

Miles, Betty. **Save the Earth, An Ecology Handbook for Kids.** New York: Alfred Knopf, 1974.

Nickelsburg, Janet. **Nature Activities for Early Childhood.** Menlo Park, CA: Addison-Wesley, 1976.

Perez, Jeannine. **Explore and Experiment.** First Teacher Press, 1988.

Richards, Roy. **Early Experiences.** London, England: MacDonald Education, 1972.

Roche, Ruth. **The Child and Science: Wondering, Exploring and Growing.** Washington, DC: Association for Childhood Education International, 1977.

Rockwell, Robert, et al. **Hug a Tree.** Mt. Rainier, MD: Gryphon House, 1983.

Russel, Helen Ross. **Ten-Minute Field Trips.** Chicago, IL: Ferguson Publishing, 1973.

Sisson, Edith. **Nature with Children of All Ages.** Englewood Cliffs, NJ: Prentice-Hall, 1982.

Skelsey, Alice, and Gloria Huckaby. **Growing Up Green.** New York: Workman Publishing, 1973.

Sprung, Barbara, et al. **What Will Happen If . . . .** Educational Equity Concepts, 440 Park Avenue South, New York, NY 10016.

Taylor, Barbara. **A Child Goes Forth.** Provo, UT: Brigham Young University, 1975.

Williams Robert, and Robert Rockwell. **Mudpies to Magnets: A Preschool Science Curriculum.** Mt. Rainier, MD: Gryphon House, 1987.

Zubrowski, Bernie. **Bubbles: A Children's Museum Activity Book.** Boston: Little, Brown and Company, 1979.

Karen Stephens is the director of the Illinois State University Child Care Center in Normal, Illinois. She is an instructor in child development for the ISU home economics department and Park Community College and currently serves as vice-president for the Midwestern AEYC.

Everybody Has One: Discovering Our Bodies

by Mary Jo Puckett Cliatt and Jean M. Shaw

Hannah and her younger sister were discussing a measle-like disease, roseola. "I feel sick. I think I have a fever. I probably have *rosemeola*," declared six year old Hannah. "It's not rosemeola! It's called *ravioli*!" cried her sister.

Four year old Chun Li said, "Teacher, wait. I've got to scratch my *armknee* (meaning elbow)!"

A five year old child ran up to the teacher on the playground and hollered, "Teacher, teacher, that mean boy said he was going to beat my guts out!" The child stopped and pondered a minute, then replied, "He can't beat my guts out. He don't even know where my guts is."

The children had been studying the five senses in kindergarten. To help the children review, the teacher asked, "Boys and girls, what have we been learning about this week?" One child eagerly responded, "The five sentences." "Can anyone name one of the five senses?" asked the teacher. "Yes," answered a little boy, "a nickel."

These illustrations show that even though young children have a limited understanding about the human body, they start to show an interest in the body and its functions at an early age. Quite often, their knowledge is a bit personalized and inaccurate. Nevertheless, they demonstrate that they can learn and want to learn various concepts related to the body.

Study of the human body increases children's awareness of the many interrelated functions of their wonderful, unique bodies. Knowledge builds appreciation of the body and the desire to care for it. Young children love learning activities that focus on and involve them. Therefore, study of the human body is perfect as a science topic for early childhood education.

Uniqueness and differences are natural parts of learning about the human body. As children learn that everyone is special, with unique characteristics and traits, their self-concepts are built. Children learn to see themselves as capable and to appreciate the capabilities of others. They can learn that everyone has both strengths and handicaps. Children learn to build on their strengths and to face their handicaps and compensate for them.

Understanding of one another's strengths and weaknesses enhances children's sensitivity and willingness to help those who need it.

Using the senses is an appropriate way for young children to learn. If we can sharpen children's abilities in using their senses, we open up whole new worlds for them. Children can take in and use more information than before. They become more sensitive to their surroundings and to other people. They build appreciation of the world around them.

Learning about the body can enhance all areas of child development. Children build understandings of body parts and body functions through concrete activities. As children converse about what they are learning; assimilate new vocabulary; and express themselves through art, music, and movement activities; they grow in their abilities to use language. Many of the activities that children participate in should encourage active physical development. They grow socially as they work together, learning to appreciate each other's uniquenesses and capabilities. Emotional growth is fostered, too. Children build confidence in their abilities as they study the body. Their natural curiosity and eagerness to learn can be maintained and heightened. Children learn to appreciate the capabilities of others as they study the body. Finally, learning activities based on the human body can promote children's creative growth because of the many inherent opportunities for divergent thought.

Unique and Different

Study of the body, especially of external features, provides numerous opportunities to teach children about similarities and differences in the body, and that *every body* has some unique features.

Classroom visitors. Invite a variety of classroom visitors over a period of time. Ask the visitors to talk about their bodies and their capabilities. For example, a parent might bring a baby and talk about his special care, tiny body parts, and the idea that the baby will grow to be big and capable (like preschoolers). The visitor could invite the children to compare their bodies and body functions to that of the baby—everyone breathes; most have two eyes, a nose, and a mouth; but the baby's movements are not as coordinated as those of the preschoolers. A grandfather could talk frankly about his body—his hair might be gray; he might be balding and wearing bifocals.

A person with a physical disability might tell the children about her limitations and ways of compensating for them. She might demonstrate any special equipment she uses and talk to the children about interacting with people with disabilities. Visitors from a variety of racial and cultural backgrounds could tell about their heritages and ways that their bodies are similar and different in appearance to those

of the children. Be sure to point out that each visitor is like the children in many ways, different from the children in some ways, and has some features that no one else has—totally unique features that make that visitor a special individual.

Body puzzles. Cut life-sized body puzzles from poster board, and let the children put them together. As the children work, encourage them to name body parts and compare the flat puzzles to their own bodies. Make sure that you use accurate colors and features to represent the

skin tones and features of various races and of any biracial children in your program.

Look at us! Let children work with partners, look at each other, and look at themselves in a mirror. Guide them to notice and discuss features such as eye color, eyelashes, teeth, and the little *bumps* on their tongues. Let the children talk about what they see and describe ways their bodies are alike and different.

Choose another body feature such as arms. Discuss the fact that most

Photograph by Roger Neugebauer

people have two arms, but some people are missing arms. Let a pair of children hold their arms side by side and look at various features—length, amount of hair, skin color, and musculature. Ask the children to show what kinds of things they can do with their arms.

People graphs. Help children categorize themselves by body characteristics and make graphs to show the results. Two simple graph formats are the sit-on graph and the felt face graph. For a sit-on graph, use plastic shelf paper. With a permanent marker, line off 50 cm. (18 inch) sections. Make removable category markers so the sit-on spaces will be reusable. For the felt face graph, cut small circles of felt using colors appropriate to the children's skin tones. Make a collar-like part for the child's name. (Figure 1 illustrates the two kinds of graphs.)

Have the children place themselves (or their markers) in the appropriate category for graphs such as "I've Lost a Tooth" or "I Haven't Lost a Tooth," "My Hair is Curly" or "My Hair is Straight," "I'm a Boy" or "I'm a Girl," "I'm Right-Handed" or "I'm Left-Handed." Discuss the graph results with the children. Do sets of people have the same number, or is one set more or less than the other? How many people marked the graph in all? What does the graph tell us?

Simulate a disability. What would it be like to have limited vision or hearing? How can a person bound in a wheelchair get along? Let children find out by helping them simulate disabilities. Use a blindfold to block vision for a short period of time. Let children wear earphones and see what it's like when sounds are muffled. Arrange for the children to take turns staying in a wheelchair for 20 to 30 minutes.

After the experiences, guide the children to talk about how they felt and how they compensated for their disabilities. Try to build appreciation for people who have permanent disabilities.

Capable Bodies

Our bodies are capable indeed. They help us move, think, and use our food. Our bodies let us receive and send many different kinds of messages. Help build children's knowledge and appreciation of how their bodies work through physical and language activities.

Sensory books. As you study the five senses, help children make

Stand On Graph

Figure 1

books about their experiences. You might use paper folded in half and devote at least one page to each sense. Label one page *My Eyes Help Me See.* Let children paste things they like to look at on the page. You could include another page labeled *Helping Me See* and let the children draw or write about their experiences looking through eyeglasses, a hand lens, or a telescope. On the page devoted to hearing, the children might paste or staple pictures of things that make noise or actual small objects such as bells, crinkled foil, or some rattly rocks in a plastic bag. Children can sprinkle spices such as curry powder or cinnamon onto glue to make a *Smelly* page. For touch, the children might glue small pieces of rough and smooth substances onto a page. For the *I Can Taste* page, let the children draw or glue pictures of favorite foods.

Special sensory walks. Take children for walks indoors and out. On each walk, focus on a different sense. You might take a walk where you look up; on another walk, you might concentrate on what you can see on the ground. On still another walk, you could take a tape recorder and make an audiotape of what familiar and strange sounds you heard. Replay the tape as you review the walk in the classroom. Take a *smelly walk* after a rain. Encourage the children to notice odors. Hang a knotted rope around an indoor area. Hang something different at each knot. Let blindfolded children feel their way along the rope and try to identify each

Photograph by Francis Wardle

*I use my **brain** to think, think, think.*
*I use my **nose** to smell.*
*I use my **eyes** to blink, blink, blink.*
*I use by **throat** to yell.*
*I use my **mouth** to giggle, giggle, giggle.*
*I use my **hips** to bump.*
*I use my **toes** to wiggle, wiggle, wiggle.*
*I use my **legs** to jump.*

Use songs and fingerplays like these to expand children's understanding of what their bodies can do for them.

Body Care

Learning to take care of our bodies is a very important aspect in the study of the human body. Young children are capable of assuming varying amounts of responsibility in caring for their bodies. They will take that responsibility more seriously if they understand the reasons for care routines.

Healthy foods/unhealthy foods. At snack time, lead children in a discussion of healthy foods. Try to help them understand something about the types of foods they need to eat every day and that they need to avoid too many unhealthy snack foods. Discuss snacks that are good for people. Be sure to serve nutri-

object. Finally, take the children on a longer walk and bring along a tasty snack. Have the children be seated and talk about each sense. What things can you see and hear? What odors are present? What can the children feel as they are seated? Now enjoy your snack with all the senses. Look at it. See if it makes a sound. Smell the snack. Let the children describe the texture of the snack. Finally, taste the snack and talk about its taste.

Body dancing. Use a large cardboard circle such as a pizza cardboard to make a spinner. Draw or paste on pictures of five or six body parts such as hands, feet, legs, torso, elbows, and head. Make a spinning

arrow in the middle. (Figure 2 shows a completed wheel.) Have a child spin the arrow, then all children must *dance* with only the body part shown on the wheel. Encourage children to try to move the specified body part in many different ways.

Songs and fingerplays. So many good songs and fingerplays are available to demonstrate the awesome capabilities of our bodies. Hap Palmer (1982) has written a song called "What a Miracle." As they sing the song, the children can demonstrate the different feats of which the body is capable. Miss Jackie Weissman (1985) has written the following body play:

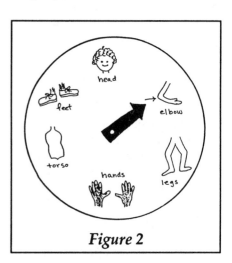

Figure 2

tious snacks. Let children cut pictures of foods from magazines and paste them on a big sheet of paper entitled "Healthy Foods" on one side and "Unhealthy Foods" on the other.

Dramatic play. Dramatic play helps young children to clarify their understandings of how to properly care for their bodies. Acting out threatening situations related to the body can help relieve children's fears of the unknown. For instance, turn your dramatic play center into a hospital. Tear up strips of sheets for bandages. Provide straight boards for splints. Make a stretcher with sheets wrapped around two long poles. Make a stethoscope from plastic tubes and funnels or use an inexpensive real stethoscope.

Add other available props. Nearby, in the library center, have several books related to the hospital. Encourage children to role play emergency and hospital care.

Another aspect of caring for our bodies is staying well groomed. Set up a beauty parlor and a barber shop in the dramatic play center. Provide props such as rollers, big bowls turned upside down for hair dryers, hair nets, cotton balls, fingernail polish bottles with colored water, safe manicure equipment, cardboard razors, shaving cream, and hand lotion. Encourage children to talk about their experiences. Extend their understanding by discussing different ways we groom our bodies at the beauty parlor and barber shop and at home.

Photograph by Francis Wardle

Felt board dolls. Make a big felt figure of the human body. You may wish to have a boy figure and a girl figure with anatomically correct features. Cut out felt grooming accessories such as toothbrushes, washcloths, Afro picks, hair brushes. Let children choose different accessories to display on the felt board next to the felt figure. Encourage the children to sing the song "This is the way we wash our bodies . . . brush our hair . . . brush our teeth . . ." depending on the accessory on the felt board.

Let the children make clothes for the felt dolls to suit different weather conditions and for different occasions such as sleeping, working, and various types of play. Provide a variety of materials for the children to use in dressing the doll. Children could make outfits from scrap fabric, lace, buttons, wallpaper, and a variety of other types of paper.

Athletic visitors. Invite an athlete to your class. Have her talk to your class about exercising. She can demonstrate and teach a few exercises to your class. Some athletes would be able to demonstrate weight lifting to the class. Encourage the athlete to discuss the importance of being careful as you exercise. Ask your visitor to discuss the importance of diet in helping to maintain a healthy body. In sports, athletes use different types of equipment to protect their bodies. If your visiting athlete uses any protective equipment, encourage him to bring the equipment and demonstrate it.

Our Insides

Children are intrigued with what's inside their bodies. They listen intently as adults discuss internal organs and often respond in egocentric ways. Teachers can help children clarify and build beginning understandings of some internal organs and their functions.

Talk through and do. Talk to children about selected internal organs and involve them as you go along. For instance, talk about the muscles and bones that are inside the hands. Ask children to wiggle the fingers of one hand and with the other hand feel the bones and muscles at work. Have the children feel the joints in their fingers and wrists. They will notice that the hand has many small bones that contribute to its flexibility. Have children examine the skin on the back of their hands and compare it to the skin on the palm. Help them notice their fingerprints and palm lines and the fact that the backs of their hands have hair while the palms do not.

Talk about breathing and the body parts that help with breathing. Have the children place their hands on their chests and breathe deeply in and out. Their lungs are about the same size as their outstretched hands, but are thicker than hands are. Have the children feel the bones that protect their lungs. Instruct them to breathe through their noses, then through their mouths. Have them breathe slowly and rapidly. Explain to children where the trachea (windpipe) is and that air goes in and out through the windpipe. Let children lay their heads on each other's backs and chests and listen to breathing with its subtle but distinctive noises.

Repeat similar activities for other internal organs.

Body shirts. Use old t-shirts to make life-sized pictures of internal organs. Check a reference book to help you with accuracy. Ask a volunteer child to don an old white t-shirt. Talk about some internal body parts such as the heart, collarbone, or stomach. As you talk, draw pictures of the organs on the shirt with a felt tip pen. Use the appropriate sizes and approximate locations. Let the children who are observing show the locations of the organs on their own bodies.

"Foot bone's connected to the ankle bone...." Use this old folk song to name bones as you work your way up the body. Have children put their hands on the bones as you name them. Play "Simon Says" with body parts. For example, "Simon says show me where your spine is," "Simon says put your elbow on your knee," or "Simon says flex a muscle."

Body clues. After exploring several internal organs, give children some clues and let them name or show where the body parts are. For instance, "This is where your food gets digested. It's like a bag where food is broken down so your body can use it" (stomach). "This body part beats all the time. It's a strong muscle. It helps blood to move around the body. Show me where it is" (heart). Help children to give other body clues for their friends to guess.

Conclusion

Study of the human body is rich in potential for learning. It is a motivating topic because it appeals to the child's egocentric nature. Activities related to the human body provide opportunities for children to grow intellectually, physically, socially, emotionally, and creatively. Teachers can use the activities suggested here or vary them and build upon them to meet the needs of children.

References

Palmer, Hap. "What a Miracle," **Walter the Waltzing Worm** (Recording No. AR555). Freeport, NY: Educational Activities, Inc., 1982.

Weissman, Miss Jackie. "I Use My Brain." Poem presented at conference of Mississippi Association on Children Under Six, Jackson, MS, 1985.

Mary Jo Puckett Cliatt and Jean M. Shaw are associate professors at the School of Education, University of Mississippi. They have extensive experience teaching and supervising teachers as they work with children. They have written articles and books on children's thinking and developmentally appropriate science and math experiences for young children.

Tyke, Tabby, and Tweety: What Children Can Learn From Pets

by Gail F. Melson

Erin watches intently as the goldfish dart up to the surface where she has just sprinkled their dinner. "They know it's time to eat!" she exclaims. Mark tries unsuccessfully to pick up the squirming class pet, a large brown rabbit named Jefferson. Sara reacts with wariness to the approach of the tomcat who has been brought to school for the day; she remembers the scratches she got when she tried to brush its hair.

Erin, Mark, and Sara, like many other preschoolers, find pet animals interesting, unpredictable creatures. They observe them, try to interact with them, and in the process learn a great deal about the nature of these animals and the care they need. Sometimes, as in Sara's case, the behavior or size of animals may provoke fears in the child, and this too may be a learning experience.

In this article, we want to explore the significance of pets for young children's development and suggest ways that teachers and parents can help this developmental process. What might exposure to pet animals teach the preschooler?

1. Learning to understand the needs of something different from the child. Preschoolers are often described as egocentric, limited to understanding the world from their own perspective. During the preschool years, children can benefit from observing and interacting with people and things with developmental needs that differ from those of the child. For example, mealtimes and types of food that are appropriate for animals are not the same as for the child. Watching a pet animal over time provides a window into the world of a creature very different from oneself.

Unlike adults or older children, animals lack the ability to adapt to the world of the preschooler. They don't politely fit into the child's games or schedule. A cat who doesn't want to be picked up, dressed in a bonnet, and wheeled around in a baby carriage will make its wishes known in no uncertain terms. In this way, animals confront the preschooler with the task of understanding different developmental needs.

2. Learning how to meet the needs of others. Taking care of a pet or watching someone else do so gives children an apprenticeship in how to nurture others. Preschoolers

have little opportunity to take care of something, except during pretend play. For young children with toddler or baby brothers or sisters, being around the baby and helping the parents take care of the baby can be valuable lessons in how to nurture others. But preschoolers' relations with their younger siblings inevitably are complicated by feelings of jealousy, envy, and resentment. Pets can be a training ground in learning about caring for others without these complex emotions. And for those preschoolers without younger brothers or sisters, pets may be an important chance for the child to be a caregiver, not just one who is cared for.

For example, we recently surveyed the parents of 140 preschoolers, about 60% of whom had at least one pet at home. We found that accord-

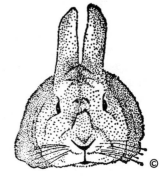

ing to parents' reports, children who had a pet but no younger brothers or sisters spent more time taking care of and playing with the pet than did preschoolers who had both a pet and a younger sibling. This suggests that pets may be an outlet for the expression of nurturance in children who lack younger brothers or sisters.

Other surveys (Kidd and Kidd) note that families with children are more likely to have a feathered or furry "family member" than are childless couples; and parents who have pets believe that they promote responsibility, empathy, and caring in their children. And "family member" is hardly an exaggeration; interviews with families who have pets (Cain) show that there are strong emotional ties to resident animals, particularly to dogs and cats.

While having pets may help all young children learn about nurturing others, it may have special benefits for boys. Girls get plenty of opportunity and encouragement to play at and practice caring for others—doll play is just one example—while boys are apt to get mixed messages. In interviews with preschoolers, we've found that by age three both boys and girls associate caring for babies with being female (Melson, Fogel, and Toda). So it's not surprising that observations of children around babies show that by the time children are five years old, boys become less interested than girls in babies and caring for them, while girls increase their interest as they mature (Melson and Fogel). Taking care of a pet, though, doesn't seem to be associated with male or female roles in children's minds. For this reason, perhaps, we've found that throughout childhood boys stay just as interested and involved in animal care as do girls. Pets may provide a special opportunity for boys to develop the nurturing side of their personalities.

3. Learning to communicate non-verbally. Unlike humans, animals communicate solely by nonverbal and nonlinguistic means. By interacting with a pet, children can learn to decode the signals that *say* hunger, play, fear, or anger. Sensitivity to nonverbal messages is an important part of being a skilled communicator, but parents and teachers pay little if any attention to developing nonverbal sensitivity in children. Pets can be great teachers of just how much can be said with a look or a movement.

4. Learning about birth, life, and death. For many young children, their first experiences with death and birth come from observing pets. When the family dog has a litter of puppies, the child may see the birth process and watch the pups nurse from the mother. This provides an excellent natural opportunity for parents to talk about the similarities and differences in birth, nursing, and early care in animals and in humans.

When the goldfish is found floating belly up in the tank or the rabbit in the toddler room dies, the child first experiences the reality of death.

Photograph by Toni Liebman

Photograph by Francis Wardle

Research (Crase and Crase) suggests that preschoolers fail to understand death as an irreversible process common to all living things; they believe that the dead can come back to life. Experiencing the death of a pet allows children to see firsthand the nature of death and the emotions that it evokes in others. The child can also see how different individuals handle these emotions. Giving children rituals, such as a funeral or burial ceremony, for marking the passage from life to death can help them understand both the irreversible nature of death and how the complex emotions aroused by death can be shared with others.

The death of a pet can also be a time when children learn about remembrance. They may see that, al-though death is final, the dead can be recalled and remain emotionally a part of one's life. Marking a pet's grave, looking at pictures of the pet, or simply talking about the animal from time to time can help the child see that the dead are not necessarily forgotten. This may help children deal with their own fears of being abandoned or forgotten by their parents or other adults.

5. The start of ecological awareness. Making contact with a pet and trying to understand it is the beginning of awareness of the many animal species in the world and the complex interdependence of all living things. Children can begin to learn respect for other animals and the importance of protecting animals and their environments. Appropriate early experiences with pets may make children more responsible toward animals in later life. Since millions of dogs, cats, and other pets are abandoned, neglected, and mistreated every year in the U.S., we need to begin developing attitudes of responsibility and protection early in life.

6. Feeling good about yourself. One adult pet owner said the great thing about her dog was: "She never talks back, she gazes up at me with love no matter what, and she always wants to play." Pets can provide children with a feeling of companionship and unconditional love. When children successfully care for a pet—giving a hungry cat its food, for example—they can experience the feeling of being needed and successfully meeting the needs of others. In other words, pets can make a preschooler feel competent and self-confident.

7. Learning responsibility. Caring for a pet is a lesson in being responsible for others. If you forget to feed an animal, it will go hungry. The consequences of not caring properly for a pet are brought home in a simple, direct way to the child. Giving preschoolers manageable care tasks, under the supervision of adults or older children, can be a valuable lesson in the importance of carrying out one's commitments to others.

8. Overcoming fears. It is not uncommon for young children to show fear of some animals, particularly large dogs and snakes. Some parents get pets to help their child overcome a fear of animals, and there is some evidence (Bowd) to support parents: a survey of parents of elementary school age children found that children with pets were less likely to be fearful of animals than were children without pets. Similar results may be true for preschoolers as well.

Some cautions. While we've been emphasizing the learning opportunities that pets provide, there are some health, safety, and developmental concerns that should be addressed. Before providing a pet for the home or school, it is important that the animal be thoroughly examined for health problems and have all necessary inoculations. Regular health care for the animal will be an ongoing part of the adult's responsibilities. (This, too, can provide a learning opportunity for the child. Accompanying a pet dog to the vet allows the child to see how doctors care for animals and to observe the emotions that a doctor visit arouses in the animal. The fear that the dog experiences at the prospect of getting a shot will be readily understood by the child!)

Not all pets are suitable for preschoolers. Surveys indicate that the most common pets for preschoolers as well as older children are dogs and cats, followed by fish, gerbils, turtles, and birds. Animals that tolerate the sometimes rough and insensitive handling of a young child are preferable to more nervous species. Check with a veterinarian or pet store concerning the breeds recommended for the younger set. For the early childhood classroom, rabbits, fish, gerbils, turtles, and snakes often are recommended.

Just as preschoolers can be rough on pets, some animals may pose risks for some children. If a child suffers from or is at risk for allergies caused by animal hairs, select short haired breeds of dogs or, to be on the safe side, stick with fish, turtles, or garden snakes. If cats or dogs are chosen, care should be taken to keep claws trimmed. Obedience training for dogs (which the child can watch) helps both dog and child by developing responsiveness and compliance in the animal. It's much easier for a small child to take a short turn

Photograph by Nancy P. Alexander

holding Rover's leash when the family is out for a walk if Rover is not trying to dash all 65 pounds of dog in the opposite direction!

Educate and supervise children concerning pet care. Parents and other adults play an important role in preparing the child to interact appropriately with the animal. Special care needs to be taken in teaching children how to handle animals that may be injured by overenthusiastic play. In addition, young children cannot be expected to remember when and how to feed the animal or to clean its living area. Preschoolers are best suited to the role of apprentices in pet care. This means that bringing a pet into the home or school "for the child(ren)" requires a considerable commitment of time and labor on the part of the adult.

Given these appropriate precautions, pets can be a valuable source

of learning for young children. What about pets in the preschool or child care center? Here are some tips.

Class pets. Having parents, grandparents, or adult friends of the child bring a family pet for a visit to the school helps the child share an important part of family life and involves the child's family in school activities. It is also a good chance to discuss the characteristics and care of animals, such as cats and dogs, that are not suitable as class pets.

Bringing pets into the school for a visit requires careful preparation. Health regulations, other children's allergies, and the welfare of the animals are some of the factors that need to be considered. It is best to schedule a pet day for a limited period, on a sunny day in a large enclosed outdoor area, and to limit the number of pets brought in.

A unit on pets. Children can bring pictures of their pets to school and talk about them during group time. All children can bring in their favorite stuffed animals. Field trips to the zoo, animal shelter, or local veterinarian can help children talk about the differences between pets and wild animals, what animals can be pets, the different foods, shelter, and care that various species need, etc. A pet store owner or zoo keeper can visit the class to talk about what animals are good as pets and how to care for them.

Cognitive skill building. Consider using materials about pets and other animals to teach classification and categorization skills. For example, a scrambled pile of pictures of different types of dogs, cats, fish, and birds can be sorted by species. A similar sorting task can help children learn the differences between farm, pet, and wild animals. Simple animal puzzles can teach children the parts of the body of common pets.

Fantasy pets. Encourage children to imagine creatures they would like for pets. Magical animals can be created by piecing together felt board pieces of animals—a zebra body, a lion head, and a mouse's tail, for example. Talk about what noises such an animal would make, how fast it would move, where it might live.

Classroom pets. Having an appropriate classroom pet can be an excellent opportunity to teach children characteristics of animals, an appreciation of their needs, and skills in caring for others. Have several children be the "special friends" of the class pet for the day, giving them responsibility (with adult supervision) for feeding the pet or cleaning the pet's living quarters. Talk about what happens to the pet after school and during

---

Recommended Books About Pets to Read with Young Children

What Pets Are Like

Allen, Marjorie and Carl. **The Marble Cake Cat.** New York: Coward, McCann and Geoghegan, 1977.

dePaola, Tomie. **The Kids' Cat Book.** New York: Holiday House, 1979.

Dunn, Judy. **The Little Duck.** New York: Random House, 1976.

Flack, Marjorie. **Angus and the Cat.** New York: Doubleday, 1931.

Floog, Jan. **Kittens Are Like That.** New York: Random House, 1976.

Floog, Jan. **Puppies Are Like That.** New York: Random House, 1975.

Henrie, Fiona. **Hamsters.** New York: Franklin Watts, 1980.

Landshoff, Ursula. **Okay, Good Dog.** New York: Harper and Row, 1978.

Langer, Nola. **Dusty.** New York: Coward, McCann and Geoghegan, Inc.

Leman, Martin. **Ten Cats and Their Tales.** New York: Holt, Rinehart and Winston, 1981.

Lord, Nancy. **My Dog and I.** New York: McGraw Hill, 1958.

Mitchell, Vanessa. **Pets and Animal Friends.** Gareth Stevens, 1985.

Robinson, Tom. **Buttons.** New York: Viking, 1938.

Skorpen, Liesel Moak. **Old Arthur.** New York: Harper and Row, 1972.

Zweifel, Frances. **Bony.** New York: Harper and Row, 1977.

About Birth and Death

Cohen, Miriam. **Jim's Dog Muffins.** New York: Greenwillow, 1984.

Stevens, Carla. **Birth of Sunset's Kittens.** Young Scott Books, 1969.

vacations to help children understand the importance of continuity of care.

In summary, bringing animals into the lives of young children in appropriate ways can yield many benefits—it can promote social sensitivity and nurturance, develop responsibility and empathy, and help children learn about the diverse animal kingdom of which they are a part.

Acknowledgement: The author wishes to thank Alan Fogel for his comments on this manuscript.

References

Bowd, A. D. "Fears and Understanding of Animals in Middle Childhood." **Journal of Genetic Psychology**, 145, 1984, 143-144.

Cain, A. O. "Pets As Family Members." **Marriage and Family Review**, 8, 1985, 3-4.

Crase, D. R., and D. Crase. "Helping Children Understand Death." **Young Children**, 32, 1976, 21-25.

Kidd, A. H., and R. M. Kidd. "Children's Attitudes Toward Their Pets." **Psychological Reports**, 57, 1985, 15-31.

Melson, G. F., and A. Fogel. "Becoming a Nurturer: How Caring Grows in Childhood." **Psychology Today**, November 1987.

Melson, G. F., A. Fogel, and S. Toda. "Children's Ideas About Infants and Their Care." **Child Development**, 57, 1986, 1519-1527.

*Gail F. Melson, Ph.D., is professor of child development and family studies at Purdue University. She is the author of **Family and Environment: An Ecosystem Perspective; Origins of Nurturance: Developmental, Biological and Cultural Perspectives on Caregiving; and Child Development: Individual, Family and Society;** as well as numerous articles on child development. She is currently conducting research on the significance of pets for children's development.*

Cooking with Preschoolers
(or Julia Child Had to Start Someplace)

by Bonnie Neugebauer

Conjure up an image of Julia Child in the kitchen—whisking, pinching, sauteing, tasting, smiling—an exciting process and a delicious, pleasing product. As I compare this image with my own laborious progression through a recipe, my efforts seem plodding, backtracking, totally dependent on the sequential directions. To give myself a bit of credit, the product of my culinary efforts is often mildy sensational; but I never feel the pride of creation, only the pride of production.

In analysis of the contrast, it becomes obvious that Julia has done a lot of "messing about" (Hawkins in Silberman, 1973) in the kitchen; I have not. The result of this opportunity for experimentation is a familiarity with materials and equipment which provides a solid base for creativity and invention beyond the scope of someone who has not messed about.

James Beard is another example of a messer:

The kitchen, reasonably enough, was the scene of my first gastronomic adventure. I was on all fours. I crawled into the vegetable bin, settled on a giant onion and ate it, skin and all. It must

have marked me for life, for I have never ceased to love the hearty flavor of raw onions (in Scharlatt, 1979).

This opportunity to investigate, taste, tear, poke, puzzle is basic to learning on many fronts. Imagine the auto mechanic who has not spent time in figuring out *what goes where* and *what happens if* and you have the mechanic who didn't quite fix it last time, and probably won't be able to next time. Or the computer programmer, the artist, the acrobat. Messing about is basic to understanding how things happen and to enjoying the process of seeing and making it happen—it is how children learn.

Creative cooking is only one of the possible directions messing about in the kitchen can take. There are also opportunities for a wealth of cognitive, social, and emotional learnings through sensory, motor, and symbolic experiences (Biber, 1971). Because the process offers such a potential for learning, cooking is *a natural* in the preschool classroom. "One can help understanding to grow in its own time and style only by offering the experiences which convey the idea" (Dorothy H. Cohen in Johnson, 1976).

Besides cognitive learning, cooking provides the child with an opportunity "to do something real" (Idyll, 1975-76)—to do something which is not merely playacting an adult role. The child can only pretend to drive a car or deliver the mail, but she can cook with real utensils and real food and actually eat the product of her efforts. And the child is working with food, something that is already very important to him.

For the preschool child, cooking satisfies the following objectives: to encourage the child to become familiar with ingredients and equipment (to know them) by experimenting with them (messing about), to expand understandings of the world (cognitive), to encourage cooperation and awareness of the needs of others (social), and to promote feelings of pride and self-confidence (emotional).

And just as messing about is wonderful for the child, it is also a key to keeping teacher interest high—adults enjoy messing about, too!

The *goal* cooking experience for the preschool child is enjoyed by the child with minimal adult intervention. This is to develop the child's feelings of autonomy and to promote a feeling of mastery. To truly cook independently, the child must be able to read; to follow the sequence of a recipe; to conserve weight, volume, and number; to understand one-to-one correspondence, fractions, time, measurement, and weight; to operate and manipulate kitchen tools and utensils; to understand cooking terms and recognize ingredients; and to plan, reverse, organize, seriate, and follow through (and probably several other skills as well).

Obviously, all of these skills are not within the scope of the preschool child. Therefore, an effort can be made to create ways to short cut the skills in order to bring independent cooking experiences within reach of the preschooler. This is done in order to facilitate "the development of an image of self as a unique and competent person" (Biber, 1971, 11). A child knows when he has accomplished something by himself.

It is necessary to make a distinction between short cutting and watering down. Simplistic experiences limit the child's potential for growth and take the developing process out of the child's control. It is the teacher who is deciding the amount of dilution. Each experience should offer the potential for growth. In other words, all the elements—equipment, ingredients, recipes—should be real so that the child can grow in perceptions and understandings—so that he can make connections in the real world. Adding water to a muffin mix robs the child of the opportunity to understand what muffins are made from. The experience must be real if it is to serve as a basis for meaning.

To explore the possibilities for cognitive learning through cooking experiences, it is helpful to focus on the High/Scope selection of key experiences (Hohmann, 1979). Basic to this curriculum approach and the work of Piaget is the concept of active learning. Piaget maintains that, in order to know an object, the child must work with it directly. Cooking offers unlimited opportunities for sensory exploration. "Both materials and ideas should be played with before their use becomes circumscribed by rules" (Biber, 1971, 14). Because the preschooler is focused on only one step of the process at a time, each part of the sequence should be an experience in and of itself.

- Select hand-operated equipment—eggbeaters, whisks—and efficient but safe utensils—knives—so that children can use them independently with minimal safety hazards.

- When children are learning to use new tools, give them time to work on the operation before you intervene.

- Select experiences appropriate for the children so that they can perform the actions. "The more the adult does, the less the children learn" (Wanamaker, 1979).

- Allow children to experiment with recipes so that they can see for themselves how each ingredient affects the product.

- At each step of the process, encourage children to taste, touch, smell, listen, observe.

Cooking provides a context for language development. As the child learns the vocabulary of cooking, she experiences greater control over her world because the words have become meaningful. Cooking easily evolves into High/Scope's plan-do-review sequence. The child helps plan and organize the project. She talks about her observations and predictions and associations as the work progresses; and she helps analyze the experience. Because the preschooler cannot reverse the sequence, the recipe can be a tool for recall.

- Encourage the child to describe what he is doing.

- Tape record the child's planning and in-process commentary as a verbal recipe that can be replayed.

- Incorporate songs and word games into the experience— "Muffin Man" song while baking muffins, "Popcorn" action song while popcorn is popping or as a recall tool.

- Create opportunities for the child to talk to her family about the experience by sending recipes or samples home.

- As children share what they have made with others, encourage them to talk about what they have done.

Focusing the child's attention on sensory interactions with materials provides the child with more information with which to represent these materials and experiences. If she has smelled, tasted, touched, and freely explored a pineapple, she will have a broader knowledge base for talking about the pineapple and for representing that pineapple through symbols. Because children come to school with many ideas about food, it can be a comfortable, rich subject for representation.

- Encourage the child to write her own recipes (or representations of the recipe) using pictures from magazines, drawings, photographs, dictation to an adult, invented writing or spelling, or other medium.

- Put empty ingredient boxes and duplicate or similar equipment in the dramatic play area to encourage the child to recreate her cooking experiences. Include copies of the recipe book.

- Encourage the child to build on the cooking experience by providing dramatic play props for grocer, farmer, baker, restauranteur.

- Include both picture and print in the recipes you provide. Leave ingredients in their original containers which usually display both picture and print identification. This supports the child's efforts to construct meaning as she makes connections between ingredient or object, picture and print.

- Label storage for equipment with both picture and print (in conventional usage of upper and lower case letters).

- Include appropriate signs— "Today's chefs are . . . ."

- Use music and movement to reenact processes and imitate— "Popcorn Song."

- Play with dough, clay, sand, water can recreate and thus strengthen the cooking experience.

As the child works on classifying objects and materials, she learns to observe and focus on attributes. Because the preschool child focuses on only one attribute at a time, materials should provide for such one-dimensional sorting. Food offers a wealth of possibilities. Long before a child can group fruits and vegetables, she can sort red apples from green apples, big pancakes from little ones. The child needs opportunities to figure out which objects belong and which objects do not belong to any given class.

- Provide materials which are different in only one way, as well as materials which are different in more than one way. This keeps the child's stage of development in mind and still allows for growth.

- Provide measuring cups which are identical except for size and measuring cups which are different colors, as well as different sizes (also other utensils such as spoons and bowls).

- Encourage the child to observe carefully—focus her attention on various attributes of each ingredient.

- Include the vocabulary which empowers the child to refine his classification skills.

Cooking offers many opportunities for seriation activities as the child learns to observe qualities and make comparisons. Having "messed about" with apples, he will be better able to seriate according to size. Having used different bowls for different purposes, he will be better able to select and store bowls according to size.

- Provide utensils in sets of three (the child in preschool can conserve three). Cups, spoons, and bowls should be stored in such a way that the child must make a match between the outline of the shape which identifies its place and the object itself. Be sure that the gradations in size are significant.

- Provide jars and pans with lids, and graduated cannisters.

- Encourage children to seriate materials by texture (hard to soft), color (light to dark), size (small to large), etc.

Counting and one-to-one correspondence matching help the child develop the concept of number. Recipes afford endless opportunities for counting—eggs, cookies, raisins, cups, children. To understand the conservation of number and amount, the child needs to work with things which must be measured (continuous materials such as flour, sugar, lemonade) and things which must be counted (discontinuous materials such as eggs, peanuts, beans). Pouring, rearranging, and distributing are all meaningful activities.

- Children can work on one-to-one correspondence by matching eggs to egg cartons or muffins to tins.

- Because it is important that the counting of the child who cannot conserve number be accepted, it is helpful to use recipes which are flexible.

- Have the children who serve samples match sample to child.

- Ingredients can be counted and poured and rearranged several times before they are actually added to the bowl.

Spatial relations and perceptions grow as children observe from various viewpoints, as they locate objects in the room, as they observe different arrangements and movements, as they manipulate objects, and as they talk about arrangements, locations, and relationships. The pre-operational child can only hold one viewpoint in his mind at a time, so he can focus only on the beginning or end. Preschoolers, therefore, need to be part of the transformation process. They need to be the agents of change, and they need every opportunity to observe the transformations taking place.

- To facilitate observations, use clear bowls and cooking pots. Use an oven with a glass panel in the door.

- Use the recipe to help the children recall the experience.

- Provide ample opportunities for work with dough. Focus observations during the stirring, shaping, baking process.

- Create, describe, compare, and eat shapes—cookies, bread, crackers, cheese.

- Provide opportunities for children to arrange food in aesthetically pleasing ways.

- Provide opportunies for children to observe changes such as those which take place in making butter, freezing popsicles, cooking applesauce, or popping popcorn (Hults, 1978).

The last key cognitive experience discussed by High/Scope involves time. For the pre-operational child, time is only subjective; meaning is derived from the child's experience and egocentric point of view. Passage of time must be related to events the child is familiar with and grows in meaning as the child experiences sequence and order in her life.

- The recipe is a natural sequencer of action. Focus on the beginning and on the end.

- Use timekeepers which make the passage visible—sand clocks, egg timers, kitchen timers.

- Use conventional time vocabulary.

- Help the child establish reference points in time.

- Establish routines and order as a framework for the child's understanding.

In addition to cognitive learning, cooking is an experience in emotional learning. As a tool in reinforcing the previously mentioned need for autonomy, repetition encourages mastery. As the child becomes more familiar with a process, he can focus on new aspects and experience greater control. Therefore, it becomes important to repeat recipes several times. One way to do this is to bake the same cake for each child's birthday.

Another is to repeat the same recipe for snack several times. During this process, allow for experimentation. The child will be more comfortable experimenting as she becomes more familiar with the recipe. If changes are initiated by the teacher, it is best to make one change at a time.

Another tool for mastery is predictability. Set up a routine for cooking that will enable the child to predict the flow. In this way, the child's energies can be directed toward the project rather than the mechanics.

- Provide a sign-up chart for chefs for the day. Cook either in partners or small groups.

- Make plans for eating, organizing, cleaning up.

Alison's Recipe for Sunshine Cookies

Beginning Cooking Lesson

The following are important points to keep in mind:

1. The child will only be able to focus on one step at a time.
2. Each ingredient should be wonderful to explore.
3. The recipe must be flexible to allow for inaccurate counting and measuring.
4. There must be potential for experimentation and growth—there should be a recipe with both pictures and print.
5. A routine should be established which will be followed throughout the year.
6. The recipe must involve action. Each child must be able to act. There must be sufficient ingredients and utensils.
7. Because the child is unable to follow the transformation process, there should be no baking involved.

Suggested first recipes/experiences:

1. Glop—peanuts, chocolate chips, raisins, sunflower seeds. Each child can select and mix and eat her own combination.
2. Squeezing orange juice. Focus is on one new skill.
3. Crunchy bananas (from Wilms, 1975). Peel bananas and cut into chunks. Roll in sesame seeds.

Next step recipes:

1. Carrot, raisin salad (more ingredients, grating skill)
2. Dip for fresh vegetables (wash, cut vegetables, and mix dip)
3. Sweet cereal balls: 1/2 cup peanut butter, 1/3 cup honey, 1/2 cup coconut, 2 cups cereal. Mix, roll into balls, and eat. (Scharlatt, 1979)

Goal Cooking Lesson

1. To foster a feeling of competency and autonomy, the goal is to enable the child to cook with little or no teacher intervention.
2. Devices in the recipe (color coding pictures and tools, depicting one step per page, including both illustrations/photos and text) allow the child who is not yet able to read conventionally to follow the instructions and sequence.
3. The product must be one the child can be proud of—enough to serve to the other children.
4. There must be room for the child's understandings to grow.
5. There must be room for the child's individualization of the recipe.

Suggested recipe:

This recipe for bread is relatively quick, and it allows for adding raisins or other innovations. It is also flexible enough to turn into bread even if the measurements are not completely accurate. It has been tested by a five year old. She needed help with the oven. Otherwise she did it all herself—and she knew she was terrific.

Candace's Pan Amasado

Mix: 2 packages yeast	Mix: 1 cup warm water	Knead.
1/4 cup warm water	1 cup milk	Shape 4 round loaves.
1/2 teaspoon sugar	2 teaspoons salt	Rest 20 minutes.
	2 tablespoons oil	Bake 40 minutes at 375 degrees.
	5 cups flour	Eat.

- Establish a method for taking turns—always go clockwise from a certain point or count to 10 (Diane Levin).

Cooking, whether in partners or groups, puts children in a situation where cooperation and interaction become necessary. In order to problem solve, the child must try to de-center—to see the other person's view. Mutually supporting patterns of interaction must be developed (Biber, 1971) if the process is to work. The child must learn to take turns, to wait, and to control his impulses.

In addition to social learning in the area of cooperation, cooking experiences can be used to explore other cultures and countries, to draw parents and professionals into the classroom, and to foster social skills.

The role of the teacher changes as the child becomes more familiar with the process and thereby more independent. In the beginning and whenever introductions of concepts and skills require it, the teacher should be an active participant in the experience. To maintain the respect element of the I-Thou-It relationship (Hawkins, 1974), the experience must be open—an exciting exploration for all. The teacher's role is one of structuring the experience so that children can figure out how and what to do (Kamii, 1978). The recipe should guide the process as children become less reliant on the teacher.

The adult is also a role model—not only on how to approach the cooking aspect of the experience but also on how and what to eat. For this reason, it is important to make good nutrition a guideline in selecting recipes.

So make is possible for the children in your program to mess about with flour and yeast, bananas and peanut butter. It will be a valuable learning experience for the Julia Childs of the future, as well as the Ronald Reagans. And remember, that it's never too late to do a bit of messing about yourself!

Recommended Resources (*) and References

Biber, Barbara, Edna Shapiro, and David Wickens. **Promoting Cognitive Growth.** New York: Bank Street College of Education, 1971.

*Burns, Marilyn. **Good for Me!** Boston: Little, Brown and Company, 1978.

*Harms, Thelma. **A Child's Cook Book.** (A Child's Cook Book, 656 Terra California Drive, #3, Walnut Creek, California 94595).

Hawkins, David. "The Triangular Relationship of Teacher, Student and Materials," from **The Informed Vision.** New York: Agathon, 1974.

Hawkins, David. "The Importance of Unguided Exploration," in Charles E. Silberman, **The Open Classroom Reader.** New York: Vintage Books, 1973.

Hohmann, Mary, Bernard Banet, and David Weikart. **Young Children in Action: A Manual for Preschool Educators.** Ypsilanti, Michigan: High/Scope, 1979.

Hults, Joanne. **Science Curriculum Guide for Preschoolers.** Boston: Boston University, August 1978.

Idyll, Janice. "Cooking with Young Children," **BAEYC Reports,** Vol. XVII, No. 2, Dec-Jan 1975-76.

*Johnson, Barbara, and Betty Plemmons. **Cup Cooking—Individual Child-Portion Picture Recipes.** Florida: Early Educators Press, 1978. Distributed by Gryphon House Inc., PO Box 275, Mt. Rainier, MD 20712.

Johnson, Georgia, and Gail Povey. **Metric Milk Shakes and Witches' Cakes.** New York: Scholastic Book Services, 1976.

Kamii, Constance, and Rheta DeVries. **Physical Knowledge in Preschool Education.** New Jersey: Prentice-Hall, Inc., 1978.

*Parents' Nursery School. **Kids Are Natural Cooks.** Boston: Houghton Mifflin Company, 1972.

Scharlatt, Elisabeth (Editor). **Kids Day In and Day Out.** New York: Simon and Schuster, 1979.

*UIC Children's Center. **I Made It Myself—A Cookbook for Young Children.** The University of Illinois at Chicago, Box 4348, Chicago, IL 60680 (312-996-8430).

*Wanamaker, Nancy, Kristin Hearn, and Sherrill Richarz. **More Than Graham Crackers.** Washington, DC: NAEYC, 1979.

*Wilms, Barbara. **Crunchy Bananas.** Salt Lake City, Utah: Sagamore Books, 1975.

A special word of thanks to Diane Levin for the class, the inspiration, and the editing.

Bonnie Neugebauer is managing editor of Child Care Information Exchange and Beginnings Books for Teachers of Young Children.

Blast Off!
Presenting Space to Preschoolers

by Sally Goldberg

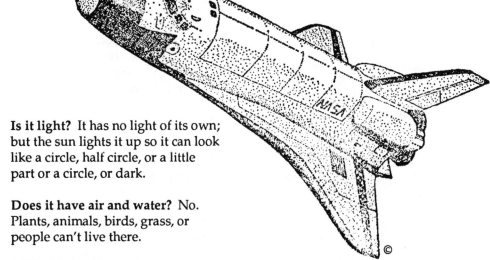

Early childhood professionals know that young children learn best by relating to their immediate real life environment, and using their five senses to see, hear, smell, touch, and taste. The real life environment of space is vast, distant, and overwhelming; but children observe the moon, sky, sun, and stars as everyday objects in their world and are intrigued by them. Children are also curious about wind and weather and relate those phenomena to the sky. They ask:

> Why is the sky blue?
> What's in the sky?
> Will the stars fall down?
> Can we go to the moon?
> What is it like on the moon?
> Where do the stars go at lunch time?

Learning about space can be fun for everyone and can readily lead to other related curriculum units. Suggestions for a preschool space unit on the **moon** follow:

Language Learning

The moon is big and round.

Where is it? It's the biggest thing in the sky at night.

Is it light? It has no light of its own; but the sun lights it up so it can look like a circle, half circle, or a little part or a circle, or dark.

Does it have air and water? No. Plants, animals, birds, grass, or people can't live there.

What would we wear there? A space suit would keep us cool in the sun and warm when we're not in the sun.

What does it look like? There are mountains, cracks, big flat places called seas, and holes called craters. We would have to walk on soft, crumbly dust and leave big footprints that would not blow away.

Does it rain/snow on the moon? The moon has no weather.

What would it feel like to be there? We would weigh less. We could pick up giant rocks and throw them,

or jump over a house. When we walked, we would go a long way and kind of float there.

How would we get there? A big rocket could take us there. The big rocket is really three small rockets: the first rocket falls away to the ground, the second rocket starts to burn fuel, and the third rocket lands on the moon.

Teaching Tools

Flannel board pictures of astronauts, rockets, etc.

A **special photo album** using magazine and newspaper photos, pictures, and postcards readily available at museums and specialty shops.

Poetry about the sky and the moon.

Excellent books about space (see reference list).

Art Activities

Life-size space suit. Draw around the children on white paper or very large paper bags or brown paper on rolls.

Binoculars. Tape toilet paper rolls together.

Rockets. Use paper rolls and ends of egg cartons and stand in the sand.

Space helmets. Cut hole for face out of large paper bags and cover hole with colored cellophane. Turn ice cream plastic buckets upside down and cut opening for face.

Sky. Sprinkle or paint with silver and white paint on black or dark blue paper.

Moon craters. Line large flat box or foam tray with plastic bag or wrap. Pour in quick set plaster of paris. While wet, drop rocks (up to 1 1/4") into plaster of paris from 6", 12", 18". Sprinkle with sand.

Music

Songs and fingerplays about moon and space.

Dance on the moon with quiet, large footsteps that carry us long distances.

Science

Space suits. Emphasize the different parts: helmet, oxygen tank, visor, pocket, glove, and boot.

Jet stream shuttle. Shape wood block on one end, curved as in a boat. Glue a cardboard or foam box on top, with one end open. Blow up balloon, hold opening, and put into box. When you release balloon opening, air will shoot out and push the boat through the water as in the spacecraft.

Parachute. Use thin nylon material or plastic wrap. Tie string or yarn to the corners, connect to piece of light wood or spool, and then throw into air.

Water vapor. Place ice cubes in jar, add water and food coloring. Cover jar. The drops of water on the outside of the jar came from water vapor in the air. The water vapor was cooled and condensed, making the air drier. In a space shuttle, the condensed water is used over and over, just as it is on earth through the lakes, clouds, and rain.

Dramatic Play

Dress-up. Supply props for dressing as astronauts—rubber gloves, helmets (bags or buckets), and turn cardboard boxes into rocket, space station, space shuttle.

Space station. We made a space station for the children to use for very short periods. We used a number of large industrial size plastic bags, taped together at several junctures. Electrical fans were used to blow up the bags and their auxiliary tunnels.

Children crawled through the tunnels and leaned on sleeping bags and pillows to view slides of the sky and stars. Because of the darkness of the inside, and the limited air supply, children were limited to five minutes at a time. As soon as they left the space station, they wanted to line up for another turn.

Footprints. Step in box lined with foil and filled with dust.

Gross Motor

Make footsteps set far apart a way to stimulate large patterns. Incorporate the idea of leaping over objects.

Photograph courtesy of Center for Early Learning & Living of the Sciences, Inc.

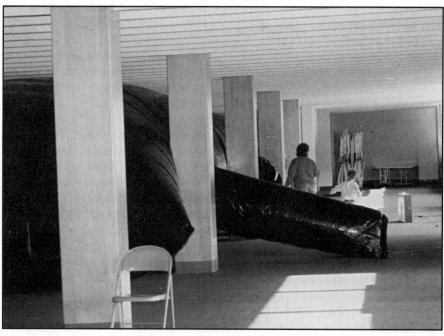

Photograph by Nancy P. Alexander

Math

Countdown for blast off.

Compare sun, earth, moon sizes with three foam balls.

Snack

Try space-like food: pellets, food in squeeze bottles, space food sticks, breakfast foods, dehydrated foods.

Environment

The key to enable young children to understand space and other related complex topics is to help them learn to ask questions: where, why, how, when, and what; to respect their questions; to work with them to seek answers; and to use the five senses approach with real objects to help relate the simple answers to the children's enlarged view of the natural world around them.

If you need a guide to begin, the Young Astronaut Program's "Living in Space," designed for preschool audiences, might be helpful as a beginning base to enhance using the children's and your own fertile imaginations.

References

Berenstain Bears on the Moon. Stan and Jan Berenstain, Random House Publishing, 1985.

Finding Out About the Sun, Moon, and Planets. Usborne Explainers, Usborne Publishers, 1981.

(A) Follett Beginning Science Book, Comets and Meteors. Issac Asimov, Follett Publishing Co., 1972.

Funk and Wagnall's New Encyclopedia of Science. Raintree Publishers, Special Projects Volume, 1986.

How and Why Wonder Book of Stars. Norman Hoss, Wonder Books, 1971.

If You Were an Astronaut. Dinah Moche, Golden Books, Western Publishers, 1985.

Let's Go to the Moon. Janis Wheat, National Geographic Society Books for Young Explorers, 1977.

Let's Read and Find Out Book, What the Moon is Like. Franklyn M. Branley, Thomas Crowell Publishing, 1963.

(The) Planets in Our Solar System. Franklyn M. Branley, Thomas Crowell Publishing, 1981.

Question and Answer Book, All About the Moon. Troll Associates, 1983.

Read About Space. James Seevers, Raintree Children's Books, 1978.

Science at Work, Projects in Space Science. Seymour Simon, Franklin Watts Publishers, 1971.

Youngest Astronauts, "Living in Space." Preschool Project, Young Astronaut Council, 1211 Connecticut Avenue NW, Suite 800, Washington, DC 20036, 1988.

Magazines

Discover
National Geographic World
Ranger Rick
Space World
Smithsonian

Sally Goldberg is executive director of the Center for Early Learning & Living of the Sciences, Inc. (CELLS). She is an educational consultant with experience as teacher, administrator, and curriculum developer.

Space Hotel by Aaron

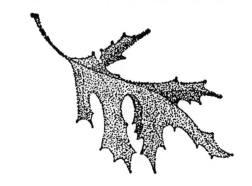

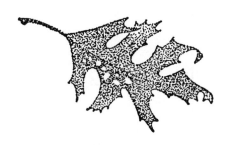

Nurturing a
Global Perspective

Who Am I?
by Felice Holman

The trees ask me,
And the sky,
And the sea asks me
Who am I?

The grass asks me,
And the sand,
And the rocks ask me
Who am I?

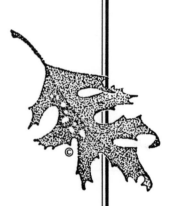

The wind tells me
At nightfall,
And the rain tells me
Someone small.

Someone small
Someone small
But a piece
of
it
all.

Lessons in Well-Being

And what of this planet of ours—what of these lives we're involved in? What is our responsibility to each and all of these children as teachers of science? For we are all teachers of science, by our curiosity or our disinterest, by the thoughtfulness of our questions, and by the poetry of our answers.

We teach science every day to the children around us, and in our interactions we teach our peers. A comment, a question, an expression of interest, or an act of wondering can affect the direction of another's thought. So here's the bottom line: we have this responsibility to live as though our actions are a blueprint for the future.

And how can we do that in a child care setting?

We can teach children to respect living things.

A sign in the flowerbox at Gentle Dragon Child Care Center in Seattle reads: "These are living things. We must respect them."

We can teach children to love and understand living things.

When Mrs. Pickering asks Sal, the program's resident cat, how his day is going, the children expect Sal to answer. They watch for body movements and listen for cat sounds that will tell them all how he feels.

We can teach children to think about their responsibilities in the world of living things.

As they leave for an outing in the woods, Ginny talks to her class about what will happen to the chipmunks if they feed them and the chipmunks come to depend on humans for food. She helps them talk about what will happen to the chipmunks when people leave this place.

We can teach children to be conscious of their impact on the environment.

Steven encourages his preschool hikers: Look around you. What signs remain of our being here? Will the next group of children be able to enjoy this place as fresh and natural?

We can teach children to do what can be done to save the planet.

Many early childhood programs recycle—saving glass and aluminum, sorting paper, turning mealtime scraps into compost. They use recycled paper. At meal and snack times, they take small portions to minimize waste.

We can teach children to be conscious of beauty, to respect beautiful things.

Lella Gandini tells us that in the schools of Reggio Emilia, Italy, ". . . you see a display of pine cones placed in order by size, and next to them a series of round, polished pebbles arranged in rows by shades of color from white to dark grey. The natural beauty of these found objects, along with their form, texture, color, and size, is highlighted by the careful attention with which they have been arranged on a lighted shelf just at children's eye level. . . . The display contains the treasures which children picked up on a special walk through the woods to the bank of the river" (**Beginnings**, Summer 1984).

We can teach children to live healthy lives.

In early childhood programs in Vancouver, where the weather is often rainy, even the babies spend significant amounts of time outdoors each day, napping under covered areas.

(By the way, who decided that rain and snow are *bad* weather?)

In many programs, exercise is part of every day. Teachers and children take many walking field trips and never ride when they can walk.

Teachers can model healthy diets by carefully choosing what they eat in front of children.

We can share wonder with children.

Rachel Carson is a master at this: "It was here that we first played our Christmas tree game. There is a fine crop of young spruces coming along and one can find seedlings of almost any size down to the length of Roger's finger. I began to point out the baby trees.

'This one must be a Christmas tree for the squirrels,' I would say. 'It's just the right height. On Christmas Eve the red squirrels come and hang little shells and cones and silver threads of lichen on it for ornaments, and then the snow falls and covers it with shining stars, and in the morning the squirrels have a beautiful Christmas tree. . . . And this one is even tinier—it must be for little bugs of some kind—and maybe this bigger one is for the rabbits or woodchucks'" (**The Sense of Wonder**, 1956).

Books to Share with Children

Spend your resources on wonder. There are thousands of quality books for young children that relate to science. Choose carefully. Look for books that flaunt wonder in glorious photographs, books that tuck wonder into verses of poetry, books that create wonder by inviting us to see the world in new ways. Treasure books that elicit responses like: "Wow!" or "Look at this!" or "I know what this is!"

Remember that our most precious resource is time, and if we spend our moments with children purusing the mediocre, we'll have no time left for the excellent. And as you read with young children, let your own sense of wonder put magic into your voice and into the moment. The following list, though not inclusive, will help get you started.

Althea. **Frogs** (illustrated by Maureen Galvani). London: Longman Group Ltd., 1977.

Ancona, George. **The Aquarium Book.** New York: Clarion Nonfiction Books, 1991.

Ancona, George. **Turtle Watch.** New York: Macmillan Publishing Company, 1987.

Arnosky, Jim. **Raccoons and Ripe Corn.** New York: Lothrop, Lee & Shepard Books, 1987.

Baker, Jeannie. **Where the Forest Meets the Sea.** New York: Greenwillow Books, 1987.

Barton, Byron. **Dinosaurs, Dinosaurs.** New York: Thomas Y. Crowell, 1989.

Barton, Byron. **I Want to Be an Astronaut.** New York: Thomas Y. Crowell, 1988.

Bash, Barbara. **Desert Giant.** Boston: Little, Brown and Company, 1989.

Bauer, Caroline Feller (editor). **Windy Day Stories and Poems** (illustrated by Dirk Zimmer). New York: J. B. Lippincott, 1988.

Bonners, Susan. **Panda.** New York: Scholastic Book Services, 1978.

Brinckloe, Julie. **Fireflies!** New York: Macmillan Publishing Company, 1985.

Brown, Margaret Wise. **Baby Animals** (illustrated by Susan Jeffers). New York: Random House, 1989.

Carle, Eric. **The Very Hungry Caterpillar.** New York: Philomel Books, 1987.

Carrick, Carol. **Big Old Bones: A Dinosaur Tale** (illustrated by Donald Carrick). New York: Houghton Mifflin Company, 1989.

Carrick, Carol and Donald. **Beach Bird.** New York: The Dial Press, 1973.

Cartwright, Sally. **Water Is Wet.** New York: Coward, McCann & Geoghegan, Inc., 1973. (Also, **Sunlight,** 1974, and **Sand,** 1974.)

Chandra, Deborah. **Who Comes?** (illustrated by Katie Lee). San

Francisco: Sierra Club Books for Children, 1995.

Cherry, Lynne (illustrator). **Grizzly Bear.** New York: E. P. Dutton, 1987.

Climo, Shirley. **King of the Birds** (illustrated by Ruth Heller). New York: Thomas Y. Crowell, 1988.

Coats, Laura Jane. **The Oak Tree.** New York: Macmillan Publishing Company, 1987.

Cole, Joanna. **How You Were Born** (photographs by Hella Hammid and others). New York: William Morrow & Company, 1984.

Cummings, E. E. **In Just-Spring** (illustrated by Heidi Goennel). Boston: Little, Brown and Company, 1973.

de Hamel, Joan. **Hemi's Pet** (illustrated by Christine Ross). Boston: Houghton Mifflin Company, 1987.

de Paola, Tomie. **The Cloud Book.** New York: Scholastic Book Services, 1975. (By the same author, **The Popcorn Book,** 1978.)

Dewey, Jennifer Owings. **At the Edge of the Pond.** Boston: Little, Brown and Company, 1987.

Dupasquier, Philippe. **Our House on the Hill.** New York: Viking Kestrel, 1988.

Edens, Cooper. **Caretakers of Wonder.** LaJolla, CA: The Green Tiger Press, 1980.

Ehlert, Lois. **Growing Vegetable Soup**. New York: Harcourt Brace Jovanovich, 1987. (By the same author, **Planting a Rainbow**, 1988.)

Ernst, Lisa Campbell. **Nattie Parsons' Good-Luck Lamb**. New York: Viking Kestrel, 1988.

Facklam, Margery. **I Go to Sleep** (illustrated by Anita Riggio). Boston: Little, Brown and Company, 1987.

Garelick, May. **Down to the Beach** (illustrated by Barbara Cooney). New York: Scholastic Book Services, 1973.

Garelick, May. **What's Inside?** (photographs by Rena Jakobsen). New York: Scholastic Book Services, 1968.

Goffstein, M. B. **Natural History**. New York: Farrar, Straus and Giroux, 1979.

Goffstein, M. B. **School of Names**. New York: Harper & Row, Publishers, 1986.

Goor, Ron and Nancy. **Heads**. New York: Atheneum, 1988.

Grabianski, Janusz. **Birds**. London: J. M. Dent & Sons Ltd., 1968.

Grillone, Lisa, and Joseph Gennaro. **Small Worlds Close Up**. New York: Crown Publishers, 1978.

Hatchett, Clint. **The Glow-in-the-Dark Night Sky Book** (illustrated by Stephen Marchesi). New York: Random House, 1988.

Hausherr, Rosmarie. **What Food Is This?** New York: Scholastic Inc. (Reading Rainbow), 1994.

Heller, Ruth. **Plants That Never Ever Bloom**. New York: Grosset & Dunlap, 1984.

Helman, Andrea. **O Is For Orca** (photographs by Art Wolfe). Hong Kong: Sasquatch Books, 1995.

Hoban, Julia. **Amy Loves the Sun** (illustrated by Lillian Hoban). New York: Harper & Row Junior Books, 1988. (By the same author, **Amy Loves the Rain**, 1989.)

Hoban, Tana. **Animal, Vegetable, or Mineral?** New York: Greenwillow Books, 1995.

Hoban, Tana. **A Children's Zoo**. New York: William Morrow & Company, 1985.

Hoban, Tana. **One Little Kitten**. New York: Scholastic Book Services, 1979.

Hofer, Angelika, and Günter Ziesler (translated by Patricia Crampton). **The Lion Family Book**. New York: North-South Books, 1995.

Horsfall, R. Bruce. **Bluebirds Seven**. Audubon Society of Portland, 5151 Northwest Cornell Road, Portland, OR 97210, 1978.

Hughes, Shirley. **Out and About**. New York: Lothrop, Lee & Shepard Books, 1988.

Imoto, Yoko. **The Picture Book of Cats**. Tokyo: Kodansha Ltd., 1984.

Johnston, Ginny, and Judy Cutchins. **Scaly Babies**. New York: Morrow Junior Books, 1988.

Kalas, Sybille. **The Goose Family Book**. Saxonville, MA: Picture Book Studio, 1986.

Kitchen, Bert. **Animal Numbers**. New York: Dial Books, 1987.

Krauss, Ruth. **The Carrot Seed** (illustrated by Crockett Johnson). New York: Harper & Row, Publishers, Inc., 1945.

Lasky, Kathryn. **The Weaver's Gift** (photographs by Christopher G. Knight). New York: Frederick Warne, 1980.

Lauber, Patricia. **The News About Dinosaurs**. New York: Bradbury Press, 1989.

Lauber, Patricia. **Seeds Pop, Stick, Glide** (photographs by Jerome Wexler). New York: Crown Publishers, 1981.

Lewis, Richard. **In the Night, Still Dark** (illustrated by Ed Young). New York: Atheneum, 1988.

Lewis, Richard. **In a Spring Garden** (illustrated by Ezra Jack Keats). New York: The Dial Press, 1965.

Lucht, Irmgard. **The Red Poppy**. New York: Hyperion Book for Children, 1994.

Macaulay, David. **The Way Things Work**. Boston: Houghton Mifflin Company, 1988.

Markle, Sandra. **Outside and Inside Snakes**. New York: Macmillan Books for Young Readers, 1995.

Martin, Bill, Jr., and John Archambault. **Listen to the Rain** (illustrated by James Endicott). New York: Henry Holt and Company, 1988.

McMillan, Bruce. **Dry or Wet?** New York: Lothrop, Lee & Shepard Books, 1988.

McNulty, Faith. **The Lady and the Spider** (illustrated by Bob Marstall). New York: Harper & Row, Publishers, 1986.

Mellonie, Bryan, and Robert Ingpen. **Lifetimes: The Beautiful Way to Explain Death to Children**. New York: Bantam Books, 1983.

Moche, Dinah L. **If You Were An Astronaut.** Racine, WI: Western Publishing Company, 1985.

Muirden, James. **Going to the Moon** (illustrated by Nigel Code). New York: Random House, 1987.

National Wildlife Federation. **COLORS In The Wild.** National Wildlife Federation, 1400 16th Street NW, Washington, DC 20036 , 1988.

Neumeyer, Peter F. **Why We Have Day and Night** (illustrated by Edward Gorey). Santa Barbara: Capra Press, 1970.

Norman, Charles. **The Hornbeam Tree and Other Poems** (illustrated by Ted Rand). New York: Henry Holt and Company, 1988.

Oechsli, Helen and Kelly. **In My Garden: A Child's Gardening Book.** New York: Macmillan Publishing Company, 1985.

Parker, Nancy Winslow, and Joan Richards Wright. **Bugs** (illustrated by Nancy Winslow Parker). New York: William Morrow & Company, 1987.

Pedersen, Judy. **The Tiny Patient.** New York: Alfred A. Knopf, 1989.

Peters, Lisa Westberg. **The Sun, the Wind and the Rain** (illustrated by Ted Rand). New York: Henry Holt and Company, 1988.

Pollock, Penny. **Water is Wet** (photographs by Barbara Beirne). New York: G. P. Putnam's Sons, 1985.

Provensen, Alice and Martin. **The Year at Maple Hill Farm.** New York: Aladdin Books, 1978.

Radin, Ruth Yaffe. **High in the Mountains** (illustrated by Ed Young). New York: Macmillan Publishing Company, 1989.

Rogow, Zack. **Oranges** (illustrated by Mary Szilagyi). New York: Orchard Books, 1988.

Romanova, Natalia. **Once There Was a Tree** (illustrated by Gennady Spirin). New York: Dial Books, 1985.

Ryden, Hope. **Wild Animals of America ABC.** New York: E. P. Dutton, 1988.

Ryder, Joanne. **Chipmunk Song** (illustrated by Lynne Cherry). New York: E. P. Dutton, 1987.

Ryder, Joanne. **The Snail's Spell** (illustrated by Lynne Cherry). New York: Viking Penguin Children's Books, 1988.

Ryder, Joanne. **Step Into the Night** (illustrated by Dennis Nolan). New York: Four Winds Press, 1988.

Ryder, Joanne. **Under the Moon** (illustrated by Cheryl Harness). New York: Random House, 1989.

Rylant, Cynthia. **This Year's Garden** (illustrated by Mary Szilagyi). New York: Macmillan Publishing Company, 1984.

Selsam, Millicent E. **Eat the Fruit, Plant the Seed** (photographs by Jerome Wexler). New York: William Morrow and Company, 1980.

Selsam, Millicent, and Joyce Hunt. **Keep Looking** (illustrated by Normand Chartier). New York: Macmillan Publishing Company, 1989.

Siebert, Diane. **Mojave** (illustrated by Wendell Minor). New York: Thomas Y. Crowell, 1988.

Simon, Seymour. **The Largest Dinosaurs** (illustrated by Pamela Carroll), New York: Macmillan Publishing Company, 1986.

Sømme, Lauritz, and Sybille Kalas (translated by Patricia Crampton). **The Penguin Family Book.** New York: North-South Books, 1995.

Spier, Peter. **Gobble, Growl, Grunt.** New York: Scholastic Book Services, 1971.

Stevens, Carla. **The Birth of Sunset's Kittens** (photographs by Leonard Stevens). New York: Young Scott Books, 1969.

Stobbs, William. **Gregory's Garden.** Oxford: Oxford University Press, 1984.

Tejima. **Woodpecker Forest.** New York: Philomel Books, 1989.

Tison, Annette, and Talus Taylor. **The Big Book of Animal Records.** New York: Grosset & Dunlap, 1985.

Tresselt, Alvin. **Hide and Seek Fog** (illustrated by Roger Duvoisin). New York: Lothrop, Lee & Shephard, 1965.

Wakefield, Pat A. (with Larry Carrara). **A Moose for Jessica** (photographs by Larry Carrara). New York: E. P. Dutton, 1987.

Waxman, Stephanie. **What Is a Girl? What Is a Boy?** New York: Thomas Y. Crowell, 1989.

Westervelt, Linda. **Roger Tory Peterson's ABC of Birds: A Book for Little Birdwatchers** (pictures by Roger Tory Peterson and Seymour Levin). New York: Universe Publishing, 1995.

Wolff, Ashley, **A Year of Birds.** New York: Viking Penguin Inc., 1984.

Ylla (story by Arthur Gregor). **Animal Babies.** New York: Scholastic Book Services, 1959.

Yoshida, Toshi. **Young Lions.** New York: Philomel Books, 1989.

Zoehfeld, Kathleen Weidner. **Seal Pup Grows Up: The Story of a Harbor Seal.** Norwalk, CT: Soundprints (Smithsonian Oceanic Collection), 1994.

Books in a Series

Animal Days (Kensington, Australia: Bay Books)

Animals From the Cold Lands.

Sharp, David. **Animals From the Hot Lands,** 1984.

Animals of the Rivers and Oceans.

Animals of Course! (New York: G. P. Putnam's Sons)

Eyes.

Feet.

Mouths.

Bailey, Jill. **Noses.**

Barron's Educational Series (Spain: Barron's Educational Series, Inc.), 1992

Julivert, Maria Angels. **The Fascinating World of Spiders** (illustrated by Marcel Socbias).

A Carolrhoda Nature Watch Book (Minneapolis: Carolrhoda Books)

Fischer-Nagel, Heiderose and Andreas. **Inside the Burrow: The Life of the Golden Hamster,** 1986.

Malnig, Anita. **Where the Waves Break Life at the Edge of the Sea** (photographs by Jeff Rotman, Alex Kerstitch, and Franklin H. Barnwell), 1985.

Schnieper, Claudia. **An Apple Tree Through the Year** (photographs by Othmar Baumli), 1987.

Eyewitness Books (New York: Alfred A. Knopf [A Darling Kindersley Book]), 1989

Bird, Butterfly & Moth, Cat, Crystal & Gem, Dinosaur, Dog, Early Humans, Elephant, Fish, Fossil, Gorilla, Horse, Insect, Jungle, Mammal, Ocean, Plant, Pond & River, Reptile, Rocks & Minerals, Seashore, Shark, Shell, Tree, Volcano & Earthquake, Weather, and **Whale.**

Eyewitness Juniors (New York: Alfred A. Knopf [Darling Kindersley Limited]), 1991

Greenaway, Frank. **Amazing Bats.**

How Your Pet Grows! (New York: Random House), 1988

Burton, Jane. **Caper the Kid.** (By the same author, **Chester the Chick, Dabble the Duckling, Freckles the Rabbit, Ginger the Kitten, Gipper the Guinea Pig,** and **Jack the Puppy.**)

Let's-Read-and-Find-Out Book (New York: Harper & Row Junior Books [Crowell/Trophy])

Aliki. **Digging Up Dinosaurs,** 1981. (By the same author, **Dinosaur Bones,** 1988; **Dinosaurs Are Different,** 1985; and **Wild and Woolly Mammoths,** 1977.)

Balestrino, Philip. **The Skeleton Inside You** (illustrated by Don Bolognese), 1971.

Berger, Melvin. **Germs Make Me Sick!** (illustrated by Marylin Hafner), 1985.

Berger, Melvin. **Switch On, Switch Off** (illustrated by Carolyn Croll), 1989.

Branley, Franklyn M. **The Beginning of the Earth** (illustrated by Giulio Maestro), 1988. (By the same author/illustrator, **Comets,** 1987.)

Branley, Franklyn M. **Eclipse: Darkness in Daytime** (illustrated by Donald Crews), 1988.

Branley, Franklyn M. **Flash, Crash, Rumble, and Roll** (illustrated by Barbara and Ed Emberley), 1985. (By the same author/illustrator, **The Moon Seems,** 1987.)

Branley, Franklyn M. **Journey into a Black Hole** (illustrated by Marc Simont), 1986. (By the same author/illustrator, **Volcanoes,** 1985.)

Branley, Franklyn M. **The Planets in Our Solar System** (illustrated by Don Madden), 1981. (By the same author/illustrator, **The Sun: Our Nearest Star,** 1988.)

Dorros, Arthur. **Feel the Wind,** 1989.

Lauber, Patricia. **Snakes Are Hunters** (illustrated by Holly Keller), 1988.

Showers, Paul. **What Happens to a Hamburger** (illustrated by Anne Rockwell), 1985.

Nature Hide & Seek (New York: Alfred A. Knopf)

Wood, John Norris. **Jungles** (illustrated by Kevin Dean and John Norris Wood), 1987.

Oceans.

A Pop-Up Field Guide (New York: Harper & Row Publishers)

Fitzsimons, Cecilia. **My First Butterflies**, 1985. (By the same author, **My First Fishes**, 1987.)

A Science Action Book (New York: Macmillan Publishing Company)

How the Weather Works.

Seymour, Peter. **Insects: A Close-Up Look** (illustrated by Jean Cassels Helmer), 1984.

A Science Activity Book (New York: Viking Penguin [Puffin Books]), 1988

The Jungles.

The Oceans.

Stick & Learn Book (New York: Harper & Row)

Stewart, Frances Todd, and Charles P. Stewart, III. **Animals and Their Environments** (illustrated by Robert Barrett), 1987. (By the same author/ illustrator, **Birds and Their Environments**, 1988.)

Stewart, Frances Todd, and Charles P. Stewart, III. **Dinosaurs and Other Creatures of Long Ago** (illustrated by Forest Rogers and Kathy Borland), 1988.

Stewart, Frances Todd, and Charles P. Stewart, III. **Fishes and Other Sea Creatures in Their Environments** (illustrated by Martin R. Koss, Jr.), 1988.

Usborne First Nature Books (London: Usborne Publishing Ltd.)

Birds.

Cox, Rosamund Kidman, and Barbara Cork. **Butterflies and Moths**, 1980.

Flowers.

Trees.

The Wonders of Nature Take-Along Library (New York: Random House [Pictureback])

Linsenmaier, Walter. **Wonders of Nature**, 1979.

McNaught, Harry. **Animal Babies**, 1977.

Singer, Arthur. **Wild Animals: From Alligator to Zebra**, 1973.

Zallinger, Peter. **Dinosaurs**, 1977. (By the same author, **Prehistoric Animals**, 1978.)

(New York: HarperCollins Publishers)

Simon, Seymour. **Sharks**, 1995.

Simon, Seymour. **Whales**, 1989. (By the same author, **Snakes, Wolves,** and **Big Cats**.)

(New York: Alfred A. Knopf [A Darling Kindersley Book]), 1995

Priddy, Roger. **Baby's Book of Nature.**

(New York: Macmillan Publishing Company [Four Winds Press])

Gibbons, Gail. **The Milk Makers**, 1985. (By the same author, **Weather Forecasting**, 1987.)

(New York: William Morrow and Company)

Arnold, Caroline. **Kangaroo** (photographs by Richard Hewett), 1987. (By the same

author/photographer, **Koala**, 1988; and **Zebra**, 1987.)

Simon, Seymour. **Jupiter**, 1985. (By the same author, **Mars**, 1987; **Saturn**, 1985; **Uranus**, 1987, and **Volcanoes**, 1988.)

(New York: William Morrow and Company [Morrow Junior Books])

Arnold, Caroline. **Rhino** (photographs by Richard Hewett), 1995.

Simon, Seymour. **Weather**, 1993. (By the same author, **Comets, Meteors,** and **Asteroids; Mountains; Neptune;** and **Star Walk**.)

(New York: Putnam Publishing Group [Philomel Books], 1985)

Yabuuchi, Masayuki. **Whose Baby?** (By the same author, **Whose Footprints?**)

(New York: G. P. Putnam's Sons)

Fischer-Nagel, Heiderose and Andreas. **A Kitten is Born.** (By the same author, **A Puppy is Born**, 1985.)

Isenbart, Hans-Heinrich. **Baby Animals on the Farm** (photographs by Ruth Rau), 1981.

Isenbart, Hans-Heinrich. **A Duckling is Born** (photographs by Othmar Baumli).

(New York: Random House), 1989

Burton, Jane. **Animals Keeping Clean.** (By the same author, **Animals Keeping Cool, Animals Keeping Safe,** and **Animals Keeping Warm**.)

Ideas, Inspiration, Answers
Resources for Teachers

The following list is just a sampling of resources available to teachers:

Allison, Linda. **The Sierra Club Summer Book.** San Francisco: Little, Brown and Company, 1989.

Ideas for warm weather explorations.

Allison, Linda. **The Wild Inside—Sierra Club's Guide to the Great Indoors.** San Francisco: Sierra Club Books, 1979.

Activities, information, ideas for exploring the indoors—including the beasts found therein.

Althouse, Rosemary. **Investigating Science With Young Children.** New York: Teachers College, 1988.

Topics are explored through 85 activities with guidelines for setting up the activity and questions to ask to help children explore, observe, compare, and talk about what is happening.

Booth, Jerry. **The Big Beast Book—Dinosaurs and How They Got That Way.** Boston: Little, Brown and Company (A Brown Paper School Book), 1988.

Excellent resource on dinosaurs.

Brown, Vinson. **Investigating Nature Through Outdoor Projects.** Harrisburg, PA: Stackpole Books, 1983.

This one is for adults who really want to get into it, who are dedicated to renewing their sense of wonder—

how to plan, initiate, and observe interactions between animals, plants, and their habitats.

Carson, Rachel. **The Sense of Wonder.** New York: Harper & Row Publishers, Inc., 1956.

Inspiration for wonder.

Cassidy, John, with David Stein. **The Unbelievable Bubble Book.** Palo Alto, CA: Klutz Press, 1987.

Lots of ideas for fun with bubbles.

Chenfeld, Mimi Brodsky. **Creative Activities for Young Children.** New York: Harcourt Brace Jovanovich, Inc., 1983.

A fabulous resource of ideas and activities, places to start, questions to ask. A boost for inspiration!

Cherry, Clare. **Creative Art for the Developing Child: Teacher's Handbook for Early Childhood Education** (Second Edition). Parsippany, NJ: Fearon Teacher Aids, 1989.

100+ projects and ideas for drawing, painting, clay play, and more.

Cohen, Richard, and Betty Phillips Tunick. **Snail Trails and Tadpole Tails: Nature Education for Young Children.** St. Paul, MN: Redleaf Press, 1993.

Eight easy-to-make mini-habitats for caterpillars, ladybugs, fish, tadpoles, earthworms, praying mantises, silkworms, and snails.

Cornell, Joseph Bharat. **Sharing Nature With Children.** Nevada City, CA: Ananda Publications, 1979.

To help children come to feel in touch with the earth and its wonders.

Cosgrove, Irene. **My Recipes Are for the Birds.** Garden City, NY: Doubleday & Company, Inc., 1975.

It would seem that birds, too, enjoy gourmet treats.

Cuffaro, Harriet K. **Experimenting with the World: John Dewey and the Early Childhood Classroom.** New York: Teachers College Press, 1995.

Dickinson, Terence. **Exploring the Sky By Day.** Ontario, Canada: Camden House Publishing, 1988. (By the same author, **Exploring the Night Sky,** 1987.)

Valuable resource for information on weather and the atmosphere—easy to understand.

Edwards, C., L. Gandini, and G. Forman (editors). **The Hundred Languages of Children: The Reggio Emilia Approach to Early Childhood Education.** Norwood, NJ: Ablex Publishing, 1993.

Fleischman, Paul. **Joyful Noise: Poems for Two Voices.** New York: Harper & Row, Publishers, 1988.

Poetry celebrating the lives of insects, written for two people to read together—perhaps two teachers performing on the subject of grass-hoppers.

Forman, George E., and Fleet Hill. **Constructive Play: Applying Piaget in the Preschool** (Revised Edition). Menlo Park, CA: Addison-Wesley Publishing Company, 1984.

100 games for learning about the physical world based on Piaget's insights.

Gallant, Roy A. **Rainbows, Mirages and Sundogs: The Sky As a Source of Wonder**. New York: Macmillan Publishing Company, 1987.

A resource on the wonders of the sky.

Gjersvik, Maryanne. **Green Fun**. Riverside, CT: The Chatham Press, Inc., 1974.

Just for fun—amusing tricks and toys to make from common plants.

Harlan, Jean, and Mary S. Rivkin. **Science Experiences for the Early Childhood Years: An Integrated Approach** (Sixth Edition). Englewood Cliffs, NJ: Prentice-Hall, Inc., 1996.

Science concepts are presented with activities (instructions and materials) and questions for discussion. Also included are suggestions for reinforcing the concept by integrating that science topic into other areas of classroom life and learning.

Herbert, Don. **Mr. Wizard's Supermarket Science**. New York: Random House, 1980.

Dramatic experiments using ordinary supplies.

Hill, Dorothy M. **Mud, Sand, and Water**. Washington, DC: National Association for the Education of Young Children, 1977.

Children need opportunities to freely explore natural materials—and exploring adults will enjoy supporting this basic need.

Holt, B. **Science with Young Children** (Revised Edition). Washington, DC: National Association for the Education of Young Children, 1989.

Kamii, Constance, and Rheta DeVries. **Physical Knowledge in Preschool Education**. Englewood Cliffs, NJ: Prentice-Hall, Inc., 1978.

Piaget's theory is the basis of this practical discussion of how early childhood teachers can impact on children's acquisition of physical knowledge. Emphasis is on children's initiative, action, and observation.

Katz, Adrienne. **Naturewatch—Exploring Nature With Your Children**. Reading, MA: Addison-Wesley Publishing Company, Inc., 1986.

Ideas for activities and explorations.

Kleinsinger, Susan Bromberg. **Learning Through Play: Science**. New York: Scholastic Inc., 1991.

Kobrin, Beverly. **Eyeopeners! How to Choose and Use Children's Books About Real People, Places, and Things**. New York: Penguin Books, 1988.

Guide to non-fiction children's books.

Kohl, Mary Ann, and Cindy Gainer. **Good Earth Art: Environmental Art for Kids**. Bellingham, WA: Bright Ring Publishing, 1991 (distributed through Independent Publishing Group, 814 North Franklin Street, Chicago, IL 60610).

200+ activities use recycled and natural materials and teach environmental responsibility.

Kohl, Mary Ann, and Jean Potter. **ScienceArts: Discovering Science Through Art Experiences**. Bellingham, WA: Bright Ring Publishing, 1993 (distributed through Independent Publishing Group, 814 North Franklin Street, Chicago, IL 60610).

Over 100 art projects that illustrate basic principles of science for children.

Kohn, Bernice. **The Beachcomber's Book**. New York: The Viking Press, 1970.

Projects using treasures from the sea.

Kohn, Judith and Herbert. **The View From the Oak**. San Francisco: Sierra Club Books, 1977.

For adults who wish to explore how different animals behave in their own habitats—how they experience space and time and how they communicate.

Macdonald Educational. **Ourselves**. Great Britain: Schools Council Publications, 1973.

Activities for exploring how our bodies work.

Martin, Laura C. **Wildflower Folklore**. Charlotte, NC: The East Woods Press, 1984.

A resource for adults who want to learn and share the history of wildflowers.

McIntyre, Margaret. **Early Childhood and Science**. Washington, DC: National Science Teachers Association, 1983.

A collection of reprinted articles from Science and Children which offers a wealth of ideas and activities for decompartmentalizing science in the classroom.

Miller, Ron, and William K. Hartmann. **The Grand Tour: A Traveler's Guide to the Solar System.** New York: Workman Publishing, 1981.

A teacher's guide to the planets (may be somewhat dated). Beautiful photographs.

Milne, Lorus J., and Margery Milne. **A Shovelful of Earth.** New York: Henry Holt and Company, 1987.

Learning to see the life beneath our feet.

Milord, Susan. **The Kid's Nature Book: 365 Indoor/Outdoor Activities and Experiences.** Charlotte, VT: Williamson Publishing, 1989.

Promotes children's awareness of the natural environment with crafts, games, stories, and poems.

Mitchell, John, and The Massachusetts Audubon Society. **The Curious Naturalist.** Englewood Cliffs, NJ: Prentice-Hall, Inc., 1980.

Crafts, games, activities, and ideas for teaching children about nature.

Moche, Dinah L. **Astronomy Today.** New York: Random House, 1986.

A teacher resource on planets, stars, space exploration.

National Science Resources Center. **Science for Children: Resources for Teachers.** Washington, DC: National Academy Press, 1988.

Nestor, William P. **Into Winter: Discovering a Season.** Boston: Houghton Mifflin Company, 1982.

A kind of curriculum web for a season—animals, birds, plants, and insects to watch for and observe.

Nickelsburg, Janet. **Nature Activities for Early Childhood.** Menlo Park, CA: Addison-Wesley Publishing Company, 1976.

Natural materials and creatures with suggestions for observation and related activities.

Ocone, Lynn. **The Youth Gardening Book.** Burlington, VT: Gardens for All, 1983.

"A good youth garden isn't just for learning the techniques of food-growing. It's also a place for mysteries and discoveries, for talking and singing, for making friends of plants, insects and fellow gardeners"—a complete guide for teachers, parents, and youth leaders.

Performanetics. **The Amusement Park for Birds,** 1994. Performanetics, 19 The Hollow, Amherst, MA 01002.

A 90 minute video of the actual teaching of a negotiated curriculum at La Villetta School in Reggio Emilia, Italy, which follows how the children design amusements for the birds who come to their playground.

Performanetics. **The Long Jump,** 1992. Performanetics, 19 The Hollow, Amherst, MA 01002.

A 120 minute video analysis of a long term project at Diana School, Reggio Emilia, Italy, showing children dealing with rules of the game, reinventing a measuring tape, to whole community involvement in a school wide athletic event directed by the children.

Petrash, Carol. **Earthways: Simple Environmental Activities for Young Children.** Mt. Rainier, MD: Gryphon House, 1992.

Categorized by season, these art and nature activities encourage the understanding and appreciation of the earth and other living things. Most activities utilize readily available materials found in nature.

Rights, Mollie. **Beastly Neighbors: All About Wild Things in the City, or Why Earwigs Make Good Mothers.** Boston: Little, Brown and Company, 1981.

A valuable affirmation that nature can be found anywhere.

Rivkin, Mary S. **The Great Outdoors: Restoring Children's Right to Play Outside.** Washington, DC: National Association for the Education of Young Children, 1995.

Rockwell, Robert E., Elizabeth A. Sherwood, and Robert A. Williams. **Hug A Tree—And Other Things To Do Outdoors With Young Children.** Mt. Rainier, MD: Gryphon House, Inc., 1983.

Learning experiences using natural environments with descriptions and follow-up suggestions.

Sheehan, Kathryn, and Mary Wardner. **Earth Child: Games, Stories, Activities, Experiments and Ideas about Living Lightly on Planet Earth.** Tulsa, OK: Council Oak Books, 1991.

Shepherd, Elizabeth. **No Bones.** New York: Macmillan Publishing Company, 1988.

A guide to such creatures as bugs, slugs, worms, ticks, spiders, and centipedes.

Sherwood, Elizabeth A., Robert A. Williams, and Robert E. Rockwell. **More Mudpies to Magnets: Science for Young Children.** Mt. Rainier, MD: Gryphon House, 1990.

126 new science projects.

Shuttlesworth, Dorothy. **Exploring Nature With Your Child.** New York: Harry N. Abrams, Inc., Publishers, 1977.

Basic information on the animal world. Beautiful photographs.

Simon, Seymour. **Pets in a Jar: Collecting and Caring for Small Wild Animals.** New York: Penguin Books, 1975.

Finding and observing wild living creatures—which are safe to capture (both for person and creature) and how to return to their natural environment.

Sisson, Edith A. **Nature With Children Of All Ages.** Englewood Cliffs, NJ: Prentice-Hall, Inc., 1982.

Activities and adventures for exploring, learning about and enjoying the natural world.

Skelsey, Alice, and Gloria Huckaby. **Growing Up Green: Children and Parents Gardening Together.** New York: Workman Publishing, 1973.

Everything you wanted to know about gardening with children.

Sprung, Barbara, Merle Froschl, and Patricia B. Campbell. **What Will Happen If . . . Young Children and The Scientific Method.** New York: Educational Equity Concepts, Inc., 1985.

A curriculum guide designed to help teachers incorporate math and science learning into classroom activities for all children. Presents several activities as carried out through the scientific method.

Stein, Sara. **The Evolution Book.** New York: Workman Publishing, 1986.

Resource for answering questions.

Stein, Sara. **The Science Book.** New York: Workman Publishing, 1979.

Interesting ideas and information leading from one question to another.

Suzuki, David. **Looking at Insects.** Toronto, Canada: Stoddart Publishing Co. Limited, 1986. (By the same author, **Looking at Plants.**)

A field trip through the insect world, with ideas for experiments.

Tilgner, Linda. **Let's Grow!— 72 Gardening Adventures With Children.** Pownal, VT: Storey Communications, Inc., 1988.

An excellent resource for gardening with children. Includes suggestions for gardening with infants, toddlers, and children with special needs.

Walpole, Brenda. **175 Science Experiments to Amuse and Amaze Your Friends.** New York: Random House, 1988.

Experiments, tricks, activities for discovery—many for school-age.

Waters, Marjorie. **The Victory Garden Kids' Book.** Boston: Houghton Mifflin Company, 1988.

An excellent handbook for beginning gardeners (both child and adult).

Whitfield, Dr. Philip, and Dr. Ruth Whitfield. **Why Do Our Bodies Stop Growing?** New York: Viking Kestrel, 1988. (Also from the Natural History Museum, **Why Do The Seasons Change?** and **Do Animals Dream?**)

Resource for answering children's questions about human anatomy.

Williams, Robert A., Robert E. Rockwell, and Elizabeth A. Sherwood. **Mudpies to Magnets—A Preschool Science Curriculum.** Mt. Rainier, MD: Gryphon House, 1987. (Also by the same authors, **Bubbles Rainbows & Worms— Science Experiments For Pre-School Children.**)

112 ready-to-use science experiments based on the natural learning process. Includes materials, vocabulary, and suggestions for continuing on.

Winnett, D., R. Rockwell, E. Sherwood, and R. Williams. **Discovery Science: Explorations for the Early Years.** Reading, MA: Addison-Wesley Publishing Company, 1996.

Over 125 science activities and ways to extend each activity into math and language.

Young Astronaut Council. **The Youngest Astronauts—Living in Space.** The Young Astronaut Council, 1211 Connecticut Avenue NW, Suite 800, Washington, DC 20036.

Series of classroom activities building on children's interest in space. A valuable resource for space-related information. Poor photographs.

Zubrowski, Bernie. **Bubbles.** Boston: Little, Brown and Company, 1979. (A Boston Children's Museum Activity Book.) Other books in the series: **Balloons, Ball-Point Pens, Blinkers and Buzzers, Clocks, Making Waves, Messing Around with Drinking Straw Construction, Messing Around with Water Pumps and Siphons, Milk Carton Blocks, Mirrors, Mobiles, Raceways, Shadow Play, Tops,** and **Wheels at Work**).

Incredible and exciting ideas for playing with bubbles.

Poetry Acknowledgments

"Grass" from **Out in the Dark and Daylight** by Aileen Fisher. Text copyright © 1980 by Aileen Fisher. Reprinted by permission of Harper & Row, Publishers, Inc.

"My Fingers" from **My Fingers Are Always Bringing Me News** by Mary O'Neill. Copyright © 1969 by Mary O'Neill. Reprinted by permission of International Creative Management, Inc.

"Some People" reprinted with permission of Macmillan Publishing Company from **Poems** by Rachel Field (New York: Macmillan, 1957).

"Water" from **Summer-Day Song** by Hilda Conkling. Copyright © 1920 and renewed 1948 by Hilda Conkling. Reprinted by permission of Random House, Inc.

"Who Am I" from **At the Top of My Voice and Other Poems** by Felice Holman. Copyright © 1970 by Felice Holman. Reprinted by permission of Valen Associates Inc.